# Studies in History and Philosophy of Science II

Scientiarum Historia et Theoria Studia, volume 3

Roberto de Andrade Martins

# Studies in History and Philosophy of Science II

Extrema
Quamcumque Editum
2021

Print edition – ISBN: 978-65-996890-3-1
E-book edition – ISBN: 978-65-996890-4-8

# Summary

# FOREWORD

This volume contains five essays in history and philosophy of science. Although they have been written several years ago, they had not been previously published, for a variety of reasons that will not be described here. No attempt was made to update them or to complement the research they describe.

(1) A priori components of science: Lavoisier and the law of conservation of mass in chemical reactions

This essay was written in 2018. A concise version was presented at the XI International Conference on History of Science and Science Education (ICHSSE), August 29-31 2018, State University of Paraiba, Brazil. At the same time, the full manuscript was circulated among the participants of the Conference, but it remained unpublished up to now. A shorter version was afterwards published, in Portuguese.[1] A related paper,[2] also addressing the *a priori* facet of the law of conservation of mass, emphasized the experimental contribution of Hans Landolt, instead of Lavoisier. Both, however, make use of Émile Meyerson's outlook on conservation laws. Many years ago, the author had already published two short papers, in

---

[1] MARTINS, Roberto de Andrade. Aspectos apriorísticos da ciência: Lavoisier e a conservação da massa nas reações químicas. Pp. 11-51, in: SILVA, Ana Paula Bispo da; MOURA, Breno Arsioli (orgs.). *Objetivos Humanísticos, Conteúdos Científicos: contribuições da história e da filosofia da ciência para o ensino de ciências*. Campina Grande: EDUEPB, 2019.

[2] MARTINS, Roberto de Andrade. Émile Meyerson and mass conservation in chemical reactions: *a priori* expectations versus experimental tests. *Foundations of Chemistry*, **21** (1): 109-124, 2019.

Portuguese, about those historical episodes and their epistemological relevance.[3]

## (2) Jevons and the role of analogies in empirical research

This paper was read at the 11th International Congress of Logic, Methodology and Philosophy of Science [Section 16. History of Logic, Methodology, and Philosophy of Science], Cracow, Poland, 20-26 August 1999. It is being published here for the first time. A former paper of mine,[4] in Portuguese, had applied Jevons' approach to the analysis of the early history of X-rays. That analysis was also included in my book on the history of radioactivity.[5]

## (3) Wave mechanics, from Louis de Broglie to Schrödinger: a comparison

This work was written for presentation at the workshop "Quantum theory: historical studies and cultural implications", held at the Universidade Federal da Paraíba, Campina Grande, Brazil, 15-17 December 2008. A shorter Portuguese version has already been issued,[6] but this full English version is now

---

[3] MARTINS, Roberto de Andrade & MARTINS, Lilian Al-Chueyr Pereira. Lavoisier e a conservação da massa. *Química Nova* **16** (3): 245-56, 1993; MARTINS, Roberto de Andrade. Os experimentos de Landolt sobre a conservação da massa. *Química Nova* **16** (5): 481-90, 1993.

[4] MARTINS, Roberto de Andrade. Jevons e o papel da analogia na arte da descoberta experimental: o caso da descoberta dos raios X e sua investigação pré-teórica. *Episteme. Filosofia e História das Ciências em Revista* **3** (6): 222-249, 1998.

[5] MARTINS, Roberto de Andrade. *Becquerel e a descoberta da radioatividade: uma análise crítica*. Campina Grande: Editora da UEPB; São Paulo: Livraria da Física, 2012.

[6] MARTINS, Roberto de Andrade. De Louis de Broglie a Erwin Schrödinger: uma comparação. Pp. 393-409, in: FREIRE JR, Olival; PESSOA JR., Osvaldo; BROMBERG, Joan Lisa (orgs). *Teoria*

published for the first time. The development Louis de Broglie's theory was dealt in detail in a former book, which was published in Portuguese.[7]

(4) The cultural relevance of astronomy in classical Antiquity
(5) The transformation of astronomical culture in the seventeenth century

These mutually complementary papers were written for presentation at the conference *Oxford Scientiae 2016: Disciplines of knowing in the early modern world* – St Anne's College, University of Oxford, 5-7 July 2016. Pondering about the title of the conference, it is easy to presume that only the second essay was intended for presentation. However, as I was writing the second paper, I felt the need of producing an account of the situation in classical Antiquity, as a preamble to the discussion of the chances that occurred in the modern period. Both parts grew uncontrollably, producing a very large paper, that was circulated during the conference. However, for publication here, I chose to split the original work in two papers, which are printed here for the first time.

**About the author**

Roberto de Andrade Martins is a Brazilian scholar. His research subjects are: foundations of physics (especially relativity theory), history of science (especially history of physics, mathematics, chemistry, astronomy and biology), and philosophy of science. He has published over 200 works on those subjects, and he has supervised over 30 dissertations and theses on his research subjects.

---

*quântica: estudos históricos e implicações culturais*. Campina Grande: EDUEPB; São Paulo: Livraria da Física, 2010.

[7] MARTINS, Roberto de Andrade; ROSA, Pedro Sérgio. *História da teoria quântica: a dualidade onda-partícula, de Einstein a De Broglie*. São Paulo: Livraria da Física, 2014.

He obtained his first degree in Physics at the University of São Paulo (USP, 1972), and a PhD in Logic and Philosophy of Science at the State University of Campinas (UNICAMP, 1987). He was a member of the faculties of the State University of Londrina (UEL), of the Federal University of Paraná (UFPR) and of the State University of Campinas (UNICAMP), where he worked throughout most of his academic career. After retiring from UNICAMP, in 2010, he was a visiting professor of the State University of Paraíba (UEPB), of the University of São Paulo at São Carlos (USP) and of the Federal University of São Carlos (UFSCar). Since 2016 he is a collaborator of the Federal University of São Paulo (UNIFESP).

He was a research fellow of the Brazilian National Council for Scientific and Technological Development (CNPq) for more than 40 years. He was president of the Brazilian Society for History of Science (SBHC) and of the South Cone Association for Philosophy and History of Science (AFHIC).

# *A PRIORI* COMPONENTS OF SCIENCE: LAVOISIER AND THE LAW OF CONSERVATION OF MASS IN CHEMICAL REACTIONS

Roberto de Andrade Martins

**Abstract**: In his book "Identité et réalité", Émile Meyerson argued for a philosophical *a priori* background in the case of all scientific conservation laws. This work discusses one of the specific cases he addressed, the conservation of mass in chemical reactions. It analyzes the attitudes of Antoine Laurent de Lavoisier and Hans Landolt regarding their own experiments concerning this law of conservation. This case study is relevant for the teaching of chemistry (and of science, in general), because it clearly shows the influence of philosophical principles on the development of science, thereby providing a nice example against the inductivist view of science.

**Keywords**: mass conservation; a priori laws; history of chemistry; philosophy of chemistry; philosophy of science; Lavoisier, Antoine-Laurent; Landolt, Hans; Meyerson, Émile

## 1. INTRODUCTION

One of the most important principles of science is the law[1] of conservation of mass (or conservation of matter). It is usually

---

[1] In this paper, "law" and "principle" will be used as equivalent terms, just for avoiding the too frequent repetition of the word "principle".

MARTINS, Roberto de Andrade. *Studies in History and Philosophy of Science II*. Extrema: Quamcumque Editum, 2021.

associated to the name of Antoine-Laurent de Lavoisier (1743-1794) or, sometimes, to that of Mikhail Vasilyevich Lomonosov (1711-1765). The usual attitude of science teachers towards this law (and to other ones) is to regard it as *true*, and to acknowledge that it was *proved*, by a series of *experiments*. Because of its eponym, it is usually thought that a single person called Lavoisier (or, perhaps, Lomonosov) was the one who proved it; and that interpretation is made more convincing by adding a specific date to the historic event.

This is the naïve view conveyed by many textbooks and popular works about the law of conservation of mass. Of course, it is completely mistaken. We know that it is impossible to provide an *experimental proof* that mass (or weight, or matter) is exactly conserved in every imaginable closed system. And the historians of science who have carefully studied the development of this law have shown that Lavoisier has never attempted to prove or to test this law – he simply stated it and used it. Even the weaker assertion that Lavoisier *discovered* the law of conservation of mass by experiments is also false.

The French philosopher of science Émile Meyerson (1859-1933) argued that this and other conservation laws have a philosophical *a priori* foundation that makes them easily acceptable; and that it was only submitted to experimental tests in the late nineteenth and early twentieth centuries. Although Meyerson's account on the principle of mass conservation was written over one century ago, in my opinion his historical description and his epistemological analysis have not been superseded. For that reason, the present article closely follows his interpretation.

This paper addresses the history and the philosophical status of the law of conservation of mass, focusing especially upon the works of Antoine-Laurent de Lavoisier and Hans Heinrich Landolt (1831-1910) – a Swiss chemist who made the most rigorous series of test of that law, around 1900. Given the scientific importance of mass conservation and the flawed exposition of its foundation in teaching, it is claimed that a more careful presentation of this episode might be a significant educational contribution.[2]

## 2. THE POPULAR VERSION: EXPERIMENTAL DEMONSTRATION

Nowadays, the Internet is a popular source of information consulted by students and teachers; and Wikipedia is one of the sites most frequently accessed by them. Some of its articles have a high quality, but that is not the case of the one concerning the conservation of mass, which contains faulty historical information:

> Historically, mass conservation was demonstrated in chemical reactions independently by Mikhail Lomonosov and later rediscovered by Antoine Lavoisier in the late 18th century. [...]
>
> By the 18th century the principle of conservation of mass during chemical reactions was widely used and was an important assumption during experiments, even before a definition was formally established, as can be seen in the works of Joseph Black, Henry Cavendish, and Jean Rey. The first to outline the principle was given by [*sic*] Mikhail Lomonosov in 1756. He demonstrated it by experiments and had discussed the principle before in 1774 [*sic*] in correspondence with Leonhard Euler, though his claim on the subject is sometimes challenged. A more refined series of experiments were [*sic*] later carried out by Antoine Lavoisier who expressed his conclusion in 1773 and popularized the principle of conservation of mass. The demonstrations of the principle led alternatives theories obsolete, like the phlogiston

---

[2] This essay was written for presentation at the XI International Conference on History of Science and Science Education (ICHSSE), August 29-31 2018, State University of Paraiba, Brazil. A shorter version, in Portuguese, was published the next year (Martins, 2019a). The English version was circulated during the Conference, but it is now being published for the first time.

theory that claimed that mass could be gained or lost in combustion and heat processes.[3]

The Wikipedia version emphasizes the idea of demonstration by experiments, either by Lomonosov[4] or by Lavoisier, stating the years when they presented the principle. Both in older and in more recent textbooks we can find similar statements concerning Lavoisier's experimental demonstration (or proof) of the law:

> Using precise balances, Lavoisier was able to demonstrate what other investigators had suspected: The quantity of matter does not change during a chemical reaction. This is the law of conservation of mass [...] (Sibring & Schaff, 1980, p. 34)

> The first experiments that conclusively demonstrated mass conservation were conducted by Antoine Lavoisier, a French chemist, just before the French Revolution. [...] Lavoisier provided the first quantitative proof of the important mass conservation principle. (Marion, 2014, p. 27)

It looks as if most of those authors have managed to reconstruct the past without the inconveniency of consulting the relevant historical sources. If one believes that scientific laws are amenable to experimental proof and if the law of mass conservation is associated to the name of Lavoisier, it seems likely that this scientist was the one who provided the experimental proof of the law.

Unfortunately, when one looks at the past through the glasses of present science and a faulty view on the nature of science,

---

[3] Wikipedia's article "Conservation of mass". Available at <https://en.wikipedia.org/wiki/Conservation_of_mass>. Accessed 17 August 2018.

[4] Lomonosov's contribution will not be discussed in this paper. Those interested in the subject may consult papers by Philip Pomper (1962) and Henry Leicester (1975). See also the book by Steven Usitalo (2013) on Lomonosov.

history inevitably becomes opaque. One only sees what he/she wants to see, not what really happened. That is the main origin of pseudohistory.

## 3. THE SEMI-POPULAR VERSION: EXPERIMENTAL DISCOVERY

Instead of *demonstration*, a few authors refer to an experimental *discovery*: "By carefully measuring the weight of each substance, Lavoisier discovered that matter is neither created nor destroyed during a chemical reaction. It may change from one form to another, but it can always be found, or accounted for" (Haven, 2007, p. 47). "Lavoisier experimentally discovered, around 1785, the law of conservation of mass" (Paty, 1999, p. 614).

It is noteworthy that the association between experiments and the *discovery* of the law of mass conservation by Lavoisier is a version accepted by several philosophers of science. In those cases, it seems that the authors are reconstructing the past from their epistemological beliefs. They know that there can be no experimental *demonstration* or *proof* of a general law from experiments; however, they do accept that, in the context of discovery, experiments may provide the inductive hint for a general law.

> The above experiments do not establish the universality claimed for $h_2$ [hypothesis 2 = mass conservation] since they concern the transactions of the particular pair mercury-oxygen: after all, the conservation of mass might be an idiosyncrasy of this couple. Lavoisier therefore tested $h_2$ in several other cases by methodically using the balance. And once he regarded $h_2$ as sufficiently corroborated – by the standards of his time – he jumped without hesitation to the general conclusion that the conservation of mass holds universally, *i.e.*, for every chemical reaction taking place in a closed system. (Bunge, 1967, pp. 256-257)

Remark that according to Mario Bunge's retrodiction or philosophical reconstruction of Lavoisier's research, the French scientist *tested* his hypothesis concerning the conservation of mass, by the *methodical* use of the balance, and *corroborated* it. Well, that is what scientists *should do*, according to Bunge's understanding of science. However, Lavoisier never actually behaved like that. Let us compare Bunge's reconstruction to Reijer Hooykaas' historical account:

> Lavoisier did not find the law of conservation of weight as a result of experiments; on the contrary he started from the metaphysical belief that nothing takes rise on its own account and – though an avowed empiricist [...] – he never performed an experiment to check the truth of this law. (Hooykaas, 1999, p. 222)

Other historians of science agree that Lavoisier accepted the conservation of weight (or mass) as an *a priori* principle and used it, but he never questioned it:

> Doubtless, Lavoisier made ample and good use of this principle [...], that is, he applied it as frequently as possible. But what is a principle? The [dictionary] *Petit Robert* says this: "Starting proposition, stated and not derived." Therefore, a principle is stated *a priori*, before the experiment. Then it is applied to the interpretation of the results and, eventually, one notices in which measure (always approximatively) the principle is followed. So, Lavoisier did not "discover" the law that bears his name after "delicate measurements" [...] but he applied it to experiments. When the experiment did not supply the expected results, he did not throw his principle to the nettles, but questioned his experience, and began again, possibly with other instruments. Thus, he did not observe anything other than what he already supposed. (Bensaude-Vincent & Journet, 1993, p. 62)

Those who are not familiar with the details of Lavoisier's researches may presume that the above description in wrong, since the French scientist is usually described as a very careful and exact experimenter. Other historians of science, however, who have devoted years of study to Lavoisier's work, have presented a devastating view of his scientific approach:

> When we follow Lavoisier's investigative pathway [...] – in particular when we reconstruct his experimental ventures at the intimate level recoverable from his laboratory notebooks – we find that he did not have a global method for ensuring that his balance sheets would balance out; that they frequently did not; that he encountered myriad errors, the sources of which he could not always identify with certainty; that he often had to calculate indirectly what he could not measure directly; that he exerted great ingenuity in the management of his data so as to make flawed experiments support his interpretations; and that he devoted much care and effort to the design of experiments so as to obviate such difficulties; but that he often settled for results he knew to be inaccurate, using his faith in the conservation principle to complete or correct the measured quantities. Much of his scientific success, I would claim, is rooted in the resourcefulness with which Lavoisier confronted the many pitfalls that lay along the quantitative investigative pathway he had chosen. (Holmes, 1982, p. 24)

Can we trust Frederic Holmes' analysis of Lavoisier's method? Of course, there is only one way of assessing his view: it is necessary to analyze Lavoisier's original account. What do the primary sources tell us?

## 4. LAVOISIER'S STATEMENT AND USE OF THE LAW

As several of other researchers of his time, Lavoisier began to *use* the principle of weight conservation without presenting any explicit statement of the same. His first publication that contained the implicit use of the principle (1770) concerned the supposed transformation of water in earth (Lavoisier, 1862, vol.

2, pp. 1-28). Weight considerations provided the main arguments he used, in that paper (Meyerson, 1908, p. 151). He also applied the same principle in his *Opuscules Physiques et Chimiques* published in 1774, where he described his first experiments on calcination (Meyerson, 1908, p. 153).

Let us illustrate Lavoisier's own presentation of the law of mass conservation. He never called much attention to the principle, and its most clear presentation appeared only in 1789, in his *Traité Élémentaire de Chimie* – not at the beginning, but on chapter 13 of the first part, where he discussed wine fermentation:

> It will be seen that, in order to arrive at the solution of these two questions, it was necessary first to be well acquainted with the analysis and the nature of the fermentable body, and the products of fermentation; for nothing is created, neither in the operations of art, nor in those of nature, and it may be advanced in principle that in every operation there is an equal quantity of matter before and after the operation; that the quality and quantity of the principles [elements] are the same, and that there are only changes, modifications.
>
> It is on this principle that the whole art of experimenting in chemistry is founded: one is compelled to assume in all [experiments] a true equality or equation between the principles [elements] of the body which one examines, and those which one extracts from it by analysis. (Lavoisier 1789, vol. 1, pp. 140-141; Lavoisier, 1796, pp. 186-187)[5]

A specific part of the above quotation is regarded as Lavoisier's statement of the principle of conservation of mass: "nothing is created, neither in the operations of art, nor in those of nature, and it may be advanced in principle that in every operation there is an equal quantity of matter before and after the operation" (Holmes, 1982, p. 24). Notice that in this

---

[5] Here I present my own translation of Lavoisier's original French, but I have also provided the reference to the corresponding page of Robert Kerr's English translation (Lavoisier, 1796).

sentence there is no mention of "mass" or "weight"; but "quantity of matter" had already been used as synonymous of mass by Isaac Newton, in the *Philosophiae Naturalis Principia Mathematica*, published one century earlier. We should also remark that Lavoisier's statement was stronger than the law of mass conservation, because he stressed the conservation of the quantity of *each particular element* in the chemical reactions. This is an additional assumption, since one might conceive that there could occur mass conservation even if there were no permanent chemical elements.

*Matériaux de la fermentation pour un quintal de sucre.*

|  | liv. | onc. | gr. | gr. |
|---|---|---|---|---|
| Eau............................................400 | » | » | » |
| Sucre.........................................100 | » | » | » |
| Levure de biere en pâte, ⎰ Eau.............. 7 | 3 | 6 | 44 |
| compoſée de ⎱ Levure ſeche.. 2 | 12 | 1 | 28 |
| TOTAL..................510 | » | » | » |

|  |  | *livre* | *once* | *gros* | *grain* |
|---|---|---|---|---|---|
| Water |  | 400 | 0 | 0 | 0 |
| Sugar |  | 100 | 0 | 0 | 0 |
| Brewer's yeast | Water | 7 | 3 | 6 | 44 |
| paste, containing | Dry yeast | 2 | 12 | 1 | 28 |
|  |  | 510 | 0 | 0 | 0 |

**Fig. 1.** Lavoisier's table of weights for the fermentation of sugar (Lavoisier 1789, vol. 1, p. 143) and the translation of its entries.

In the chapter on fermentation, Lavoisier presented the quantitative details of his experiments. To understand his measurements, it is necessary to know the units of weight he used (Partington, 1961-1970, vol. 3, p. 377). The basic unit was the old French pound (*livre*, or *pois de marc*). We know that it amounted to 489.5058 g. Its subdivisions were: 1/16 of the *livre* was named *once* (30.5941 g); 1/8 of the *once* was named *gros* (3.8242 g); and 1/72 of the *gros* was called *grain* (0.0531 g, or 53.1 mg). So, one *livre* contained 9,216 *grains*.

The substances he used for fermentation were water, sugar and yeast; their amounts, according to Lavoisier, were 400 *livres* of water, 100 *livres* of sugar, and 10 *livres* of brewer's yeast paste; and the latter contained 7 *livres*, 3 *onces*, 6 *gros* and 44 *grains* of water and 2 *livres*, 12 *onces*, 1 *gros* and 28 *grains* of dry yeast (Lavoisier 1789, vol. 1, p. 143; see Fig. 1). Notice that his quantitative data suggest a precision of one grain (about 1/20 g) in his weighing.

Were his balances really so exact? We know that his two most accurate instruments were a balance produced by Pierre-Bernard Mégnié that could detect 5 mg for a total weight of 600 g (about one part in 100,000); and another one produced by Nicolas Fortin, which could detect about 25 mg when charged with 10 kg (about two parts in one million). These were the best instruments available at that time (Bensaude-Vincent & Journet, 1993, p. 49). In the fermentation experiments, the total weight was 510 *livres* (without computing the vessels containing the substances), corresponding to about 250 kg. If he could measure those quantities with the accuracy of one *grain* (0.0531 g), then his precision would be about two parts in ten million. However, his best balances could not attain such exactitude.

Lavoisier did not really make his experiments using the very large weight of materials described in his table. Instead of 100 *livres* of sugar, he acknowledged that he used just a few pounds:

> In these results I have carried the precision of calculation to the grain. However, it is not possible yet to carry this kind of experiment to such great exactness. But as I have worked only on a few pounds of sugar – and, in order to make comparisons, I have been obliged to reduce them to the *quintal* [one hundred *livres*] –, I have deemed it my duty to keep the fractions such as the computation had given to me. (Lavoisier 1789, vol. 1, pp. 148-149; Lavoisier, 1796, pp. 194-195)

Therefore, Lavoisier's tables do not provide his actual measurements. They show the proportional values he computed, corresponding to a much larger hypothetical amount of materials.

*Détail des principes constituans des matériaux de la fermentation.*

```
liv.   onc. gr. grains.                              liv. onc. gr.  grains.
407    3    6   44  d'eau ⎧Hydrogène.....  61   1   2   71,40
       composées de      ⎩Oxygène.......346   2   3   44,60

                         ⎧Hydrogène....   8   »   »     »
100 l. de fucre compo-   ⎨Oxygène......  64   »   »     »
       fées de           ⎩Carbone......  28   »   »     »

liv. onc. gr.  gr.       ⎧Carbone......  »  12   4   59,00
  2   12   1   28 de le- ⎨Azote........  »   »   5    2,94
 vure feche compofées de ⎨Hydrogène....  »   4   5    9,30
                         ⎩Oxygène......  1  10   2   28,76
                                         ________________________

           TOTAL............510   »   »     »
```

| | | livre | once | gros | grain |
|---|---|---|---|---|---|
| 407 liv. 3 onc. 6 gr. 44 grains water | Hydrogen | 61 | 1 | 2 | 71.40 |
| | Oxygen | 346 | 2 | 3 | 44.60 |
| 100 liv. of sugar, containing | Hydrogen | 8 | 0 | 0 | 0 |
| | Oxygen | 64 | 0 | 0 | 0 |
| | Carbon | 28 | 0 | 0 | 0 |
| 2 liv. 12 onc. 1 gr. 28 grains of dry yeast, containing | Carbon | 0 | 12 | 4 | 59.00 |
| | Azote | 0 | 0 | 5 | 2.94 |
| | Hydrogen | 0 | 4 | 5 | 9.30 |
| | Oxygen | 1 | 10 | 2 | 28.76 |
| | | 510 | 0 | 0 | 0 |

Fig. 2. Lavoisier's table of weights of the several elements contained in the reacting substances used in the fermentation of sugar (Lavoisier 1789, vol. 1, p. 144) and the translation of its entries.

Lavoisier also provided the detailed elementary composition of the materials he used, giving the weights of hydrogen, oxygen, carbon, and nitrogen ("azote") contained in each substance (Fig. 2). Of course, the amount of each element was not *measured*: it was computed taking into account Lavoisier's estimations of the components of each substance, in some previous researches he had made. In the case of sugar, for instance, Lavoisier stated that the proportions of the elements were approximately (*à-peu-près*) the following: Hydrogen 8%, Oxygen 64% and Carbon 28% (Lavoisier 1789, vol. 1, p. 142). Notice that he used those proportions as if they were exact, and that in his table he recorded fractions of hundredths of grain (less than one milligram), which were impossible to measure.

According to Lavoisier, the products of fermentation were these (using the names he used):

|  | *livre* | *once* | *gros* | *grain* |
|---|---|---|---|---|
| Carbonic acid | 35 | 5 | 4 | 19 |
| Water | 408 | 15 | 5 | 14 |
| Dry alcohol | 57 | 11 | 1 | 58 |
| Dry acetous acid[6] | 2 | 7 | 8 | 0 |
| Sugar residue | 4 | 1 | 4 | 3 |
| Dry yeast | 1 | 6 | 0 | 50 |
|  | 510 | 0 | 0 | 0 |

According to Lavoisier's data, there was an astonishingly exact match of the total weights of the products and of the reacting substances. Not a single grain was lost or acquired. Anyone who has really made laboratory measurements will be unable to believe that he could obtain this result. It is very likely that in the process of drying the several substances he could not track down the exact amount of water that they lost; and the total

---

[6] There is a typographical mistake in Lavoisier's book, where the number of *onces* is omitted.

weight of water had to be adjusted, so that the total weight would obey the law of conservation of mass.

> The numerical results of his [Lavoisier's] experiments often are too good to be true: more modern and more precise methods would not yield such marvelous results as he mentions for his experiments on fermentation. (Hooykaas, 1999, p. 222)

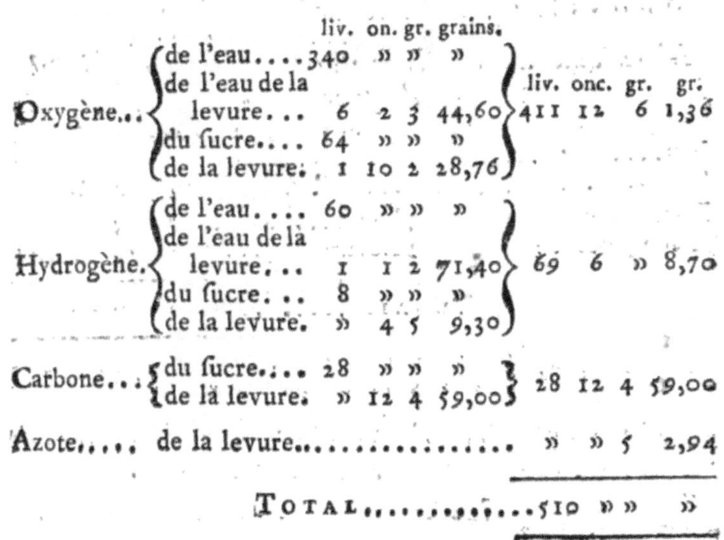

|  | *livre* | *once* | *gros* | *grain* |
|---|---|---|---|---|
| Oxygen | 411 | 12 | 6 | 1.36 |
| Hydrogen | 69 | 6 | 0 | 8.70 |
| Carbon | 28 | 12 | 4 | 59.00 |
| Azote |  |  | 5 | 2.94 |
| Total | 510 | 0 | 0 | 0 |

**Fig. 3**. Lavoisier's table showing the weights of the several elements contained in the substances used in the fermentation experiment (Lavoisier 1789, vol. 1, p. 144).

Lavoisier also provided the amounts of each element of the substances he used, both before and after fermentation. According to his account, the composition of the substances used in the fermentation experiment was that shown in Fig. 3.

*RÉCAPITULATION des résultats obtenus par la fermentation.*

|  |  | liv. | on. | gr. | gr. |
|---|---|---|---|---|---|
| **409 10 » 54 d'oxygène.** | de l'eau............ | 347 | 10 | » | 59 |
|  | de l'acide carbonique. | 25 | 7 | 1 | 34 |
|  | de l'alkool.......... | 31 | 6 | 1 | 64 |
|  | de l'acide acéteux... | 1 | 11 | 4 | » |
|  | du résidu sucré....... | 2 | 9 | 7 | 27 |
|  | de la levure......... |  | 13 | 1 | 14 |
| **28 12 5 59 de carbone.** | de l'acide carbonique. | 9 | 14 | 2 | 57 |
|  | de l'alkool.......... | 16 | 11 | 5 | 63 |
|  | de l'acide acéteux... |  | 10 | » | » |
|  | du résidu sucré...... | 1 | 2 | 2 | 53 |
|  | de la levure......... |  | 6 | 2 | 30 |
| **71 8 6 66 d'hydrogène.** | de l'eau............ | 61 | 5 | 4 | 27 |
|  | de l'eau de l'alkool.. | 5 | 8 | 5 | 3 |
|  | combiné avec le carbone dans l'alkool.. | 4 | » | 5 | ». |
|  | de l'acide acéteux... |  | 2 | 4 | » |
|  | du résidu sucré...... |  | 5 | 1 | 67 |
|  | de la levure......... |  | 2 | 2 | 41 |
| **2 37 d'azote**.................... |  |  |  | 2 | 37 |
| **510 » » »** |  | 510 | » | » | » |

|  | *livre* | *once* | *gros* | *grain* |
|---|---|---|---|---|
| Oxygen | 409 | 10 | 0 | 54 |
| Hydrogen | 71 | 8 | 6 | 66 |
| Carbon | 28 | 12 | 5 | 59 |
| Azote |  |  | 2 | 37 |
| Total | 510 | 0 | 0 | 0 |

**Fig. 4**. Lavoisier's table of the weights of each element contained in the products of fermentation (Lavoisier 1789, vol. 1, p. 148).

According to Lavoisier's data, the amounts of each element obtained in the products of the fermentation experiment were different from the initial weights, although the total weight was the same (Fig. 4).

Lavoisier did not attempt to explain the considerable differences between the initial and final weights of the elements, which amounted to about 2 *livres* in the cases of Oxygen and Hydrogen. Transforming into kilograms and rounding off his figures, the initial mass of Oxygen was 201.577 kg and after fermentation it was only 200.516 kg; that of Hydrogen was initially 33.980 kg and it increased to 35.026 kg. Following a strictly empiricist point of view, the experiment *refuted* the conservation of mass of each element, in fermentation.

We must understand that the amount of each element was not *measured*, it was computed from the analysis of each product according to previous experiments. The lack of agreement shows that Lavoisier's analysis of each substance had significant errors, of the order of 3 to 5%.

At the end of the same chapter of his book, Lavoisier stated:

> I shall finish what I have to say about vinous fermentation, by observing that it may provide a means of analyzing sugar and every vegetable fermentable matter. Indeed, as I have already pointed out at the beginning of this article, I may consider the substances submitted to fermentation, and the result obtained after fermentation, as an algebraic equation; and supposing successively each element of this equation as unknown, I can obtain from it a value, and thus rectify experience by calculation and calculation by experience. I have often taken advantage of this method to correct the first results of my experiments, and to guide me concerning the precautions that should be taken for repeating them [...] (Lavoisier 1789, vol. 1, p. 151; Lavoisier, 1796, p. 197)

So, Lavoisier recognized that he used the law of mass conservation to *correct* the results of his experiments. He adjusted the quantities so that there was an exact match between the sum of the masses of the reacting substances and that of their products. The data he published was not what he measured, it was what the experiment *should have shown*, according to the principle of conservation of mass.

> In Lavoisier's publications the numerical values of the quantities of substances always agree 100% with the yields of reactions found for their component elements. This is too good to be true and he admitted this himself when he said that he "corrected experience by calculation". (Hooykaas, 1999, p. 223)

Whenever a disagreement occurred, this was never regarded by him as a *refutation* of the law of mass conservation, but the anomaly could conduct him to take additional precautions in further trials, because the lack of concordance was interpreted by him as some mistake that occurred *in the experiments*.

## 5. LAVOISIER AND THE CONSERVATION OF CALORIC

It is remarkable that Lavoisier's thought was frequently guided by conservation hypotheses, in several fields – not only in chemistry and physics:

> There is therefore, at least for most of the territorial productions of the kingdom of France, an equation, an equality between what is produced and consumed; thus, to know what is produced, it is enough to know what is consumed [...] (Lavoisier, 1791, p. 9)

It is well known that Lavoisier died in the guillotine, during the French revolution, because of his involvement in tax collecting – not because of his scientific works. In 1770, at the age of 26, Lavoisier had acquired a share in the French "Ferme générale", a company which collected taxes for the king and distributed a bulky part of the received money between its members (Aykroyd, 1935, pp. 12-17). His position as a *fermier général* and other jobs he obtained because of that involvement, such as the supervision of tobacco and gunpowder production (Scheler, 1973), were the main sources of his personal income. He kept detailed accounts of finances, registering the amounts of incoming and outgoing money and the resulting balance. Of course, money does not usually disappear, nor is it created from nothing – it is conserved in financial transactions. Lavoisier's financial book-keeping was similar to his computations concerning the conservation of matter:

> His [Lavoisier's] training in keeping accounts and preparing balance-sheets as a *Fermier Général* influenced his scientific work, and there is a close resemblance in form between his official memoranda (printed in the *Oeuvres*, especially in Vol. VI) and his scientific memoirs. (Partington, 1961-1970, vol. 3, p. 376)

> [...] nothing is created, either in the operations of art, or in those of nature, and one can state as a principle that in every operation there is an equal quantity of material before and after the operation. It is recognized that this statement was the operating principle on which Lavoisier based his "balance sheet" method of experimentation [...] (Holmes, 1982, p. 24)

> [...] Lavoisier's scales were more than a precision instrument. They materialized an intellectual strategy of balancing inputs and outputs that Lavoisier used daily in his book-keeping activity as a tax collector and also in his reflections on rational economics — both domestic, and national. (Bensaude-Vincent & Simon, 2008, p. 86)

It is not widely known that, besides the principle of conservation of weight, Lavoisier also accepted another conservation law which was later rejected: that of quantity of heat (or caloric).

In the seventeenth and eighteenth centuries, some authors regarded heat as produced by microscopic (invisible) motion, and other supposed that it was a substance. Francis Bacon, René

Descartes, Robert Boyle and Daniel Bernoulli had endorsed the view of heat-motion; Pierre Gassendi, Leonhard Euler, Herman Boerhaave, Georges-Louis Leclerc (comte de Buffon), Joseph Black, Jean-André De Luc and Johan Carl Wilcke supported the heat-substance concept (Barnett, 1946; Boyer, 1943; Morris, 1972, p. 28). Calorimetric concepts and experiments, developed in the second half of the eighteenth century, were connected to the assumption of heat as a substance, and after the works of Wilcke, De Luc and Black,

> [...] the triumph of the first of these theories seemed complete. From then on, heat is treated as a real substance that passes from one body to another, without any change of its quantity (which one had learned to measure) and that, if it ceases to be manifested to our sensation and if the thermometer does not detect it, none the less continues to exist in a particular state. (Meyerson, 1908, p. 173)

Boerhaave attempted to measure the weight of heat and he was unable to find any definite change of weight when a body was heated, concluding as a result that heat was an imponderable substance (Boyer, 1943, p. 448) – an inference that was later accepted by Lavoisier.

> The notion of imponderable substances was well established during the eighteenth century. Electricity was generally understood to consist of one or two imponderable fluids. Magnetism could be similarly explained. So could light and heat. Lavoisier extended the explanatory model by making two imponderable fluids, light and caloric, simple substances for the chemist. (Hooykaas, 1999, p. 74)

Lavoisier came to accept the idea of a heat-substance, which he called by several names, such as "igneous fluid", "matter of heat" and finally "caloric" from 1787 onwards (Morris, 1972, p.

2)[7]. In his chemistry, caloric was one of the elementary substances (Lavoisier, 1789, vol. 1, p. 192; Siegfried, 1982, p. 33), and the first chapter of his *Traité Élémentaire de Chimie* is devoted to its presentation. According to Lavoisier, caloric is not heat, it is its cause (Lavoisier, 1789, vol. 1, pp. 4-5).

*TABLEAU DES SUBSTANCES SIMPLES.*

| | Noms nouveaux. | Noms anciens correspondans. |
|---|---|---|
| *Substances simples qui appartiennent aux trois règnes & qu'on peut regarder comme les élémens des corps.* | Lumière........... | Lumière. |
| | Calorique......... | Chaleur. / Principe de la chaleur. / Fluide igné. / Feu. / Matière du feu & de la chaleur. |
| | Oxygène.......... | Air déphlogistiqué. / Air empiréal. / Air vital. / Base de l'air vital. |
| | Azote............ | Gaz phlogistiqué. / Mofete. / Base de la mofete. |
| | Hydrogène....... | Gaz inflammable. / Base du gaz inflammable. |

**Fig. 5.** The beginning of Lavoisier's table of simple substances (elements), which included light (*lumière*) and caloric (*calorique*), together with oxygen, nitrogen and hydrogen (Lavoisier, 1789, vol. 1, p. 192).

According to Lavoisier, gases are produced by the chemical combination of a material substance with caloric, and oxygen gas is a specific instance of such a combination (Meyerson, 1908, p. 151; Guerlac, 1976, p. 220)[8]. For that reason, although

---

[7] The name "caloric" has been applied for the first time in the book *Méthode de nomenclature chimique,* jointly authored by Guyton de Morveau, Lavoisier, Bertholet and Fourcroy (Guyton de Morveau *et al.*, 1787, p. 78). There, caloric was described as a chemical element, an idea that was later retained by Lavoisier.

[8] Lavoisier first published this idea in his 1777 paper, "De la combinaison de la matière du feu avec les fluides évaporables: et de

*oxygen* is a simple substance (element), it is possible to decompose *oxygen gas*. This is a topic dealt with by Lavoisier in the fifth chapter of his *Traité*, "The decomposition of oxygen gas by sulfur, phosphorus, and coal, and the formation of acids in general", where we find descriptions such as this:

> This experiment clearly proves that at a certain temperature, oxygen has more affinity with phosphorus than with caloric; that, consequently, phosphorus decomposes the oxygen gas, that it seizes its base, and that the caloric, which becomes free, escapes and dissipates by distributing itself among the surrounding bodies. (Lavoisier, 1789, vol. 1, p. 60)

Amounts of ponderable matter can be assessed with a balance. Caloric, however, is an imponderable substance, it cannot be weighed – but its quantity can be ascertained by calorimetric experiments. Following the pioneering ideas and experimental researchers of Joseph Black and other previous authors, Lavoisier and Pierre-Simon de Laplace studied phenomena of heat transference between bodies by means of a very sophisticated ice calorimeter. They also investigated the heat generated by chemical reactions and by living bodies (Guerlac, 1976). In their joint paper published in 1783, describing the instrument and its use (Lavoisier, 1862, vol. 2, pp. 283-333), Lavoisier and Laplace mentioned the two opposing theories of heat.

> *The quantity of free heat remains the same in the simple mixture of bodies*. This is evident if heat is a fluid which tends to equilibrium; and if it is simply the living force which results

---

la formation des fluides élastiques aeriformes" (Lavoisier, 1862, vol. 2, pp. 212-224). There is a full and commented English translation (Best, 2015-2016) of another of Lavoisier's early works on the subject, "Réflexions sur le phlogistique, pour servir de suite à la théorie de la combustion et de la calcination" (Lavoisier, 1862, vol. 2, pp. 623-655).

> from the internal motion of matter, the principle in question is a consequence of that of the conservation of living forces. The conservation of free heat, in the simple mixture of bodies, is therefore independent of any hypothesis as to the nature of the heat; it has been generally accepted by physicists, and we will adopt it in the following researches. (Lavoisier, 1862, vol. 2, p. 287)[9]

Laplace did prefer the heat-motion hypothesis and Lavoisier always adopted the heat-substance concept (Guerlac, 1976, pp. 244-245), but they reached a compromise and stated that the experimental investigations they presented were independent of the particular interpretation adopted (Morris, 1972, pp. 30-31). In all those researches, they were guided by the idea that heat is conserved: it may be hidden in bodies, but it cannot be created or destroyed.

Independently of the particular justification, the law of conservation of heat was accepted by all researchers who did calorimetric experiments. Johan Carl Wilcke, Jean-André De Luc, and Joseph Black accepted the indestructibility of the matter of heat, before Lavoisier (Meyerson, 1908, p. 191). The conservation of heat was *suggested*, *accepted* and *applied* as a reliable law, just as in the case of the law of conservation of matter. We know that the law of conservation of heat was flawed. We accept that *energy* is conserved; heat is one of the many forms of energy and it can be created or destroyed, when there are energy transformations. Heat itself is not conserved, except in some very special phenomena – those that only involve heat conduction between inert bodies.

## 6. WAS LAVOISIER'S LAW AN APRIORISTIC TRUTH?

It is clear that in the fermentation experiment Lavoisier was *assuming* and *using* the law of conservation of mass, but he was

---

[9] "Free heat" was understood by Lavoisier as the part of caloric that was not chemically combined with material substances.

not *testing* it. Indeed, historians of science have never found a single experiment in his published works or unpublished notes that could be interpreted as a test of that law.

> Lavoisier did not find the law of conservation of weight as a result of experiments; on the contrary he started from the metaphysical belief that nothing takes rise on its own account and – though an avowed empiricist [...] he never performed an experiment to check the truth of this law. (Hooykaas, 1999, p. 222)

He frequently *applied* the law, long before he explicitly stated it. According to Holmes, "His first notable experiments on the transmutation of water in 1768-70 relied on that method, and it pervaded all of his experimental investigations through the next two decades" (Holmes, 1982, p. 24).

> Lavoisier's applications of the law of the conservation of mass were inspired by his faith in an incontrovertible truth. Thus, in his early publications he did not weigh all the substances taking part in a reaction, but commonly used the law to weigh them indirectly. (Hooykaas, 1999, p. 222)

Consulting several of Lavoisier's works, it is possible to notice that he regarded the conservation of mass as an obvious statement and similar to the mathematical equality between two terms and their sum. In his memoir on the decomposition of water (1781)[10], after describing the burning of hydrogen and oxygen, Lavoisier affirmed:

> [...] we could not ascertain the exact quantity of the two airs with which we had thus made the combustion; but, as it

---

[10] The full work is reproduced in the second volume of Lavoisier's *Oeuvres* (Lavoisier, 1862, pp. 334-359): "Mémoire dans lequel on a pour objet de prouver que l'eau n'est point une substance simple, un élément proprement dit, mais qu'elle est susceptible de décomposition et de recomposition" (1781).

is no less true in physics than in geometry that the whole is equal to its parts, from the fact that we had obtained only pure water in this experiment, without any other residue, we thought that we were justified in concluding that the weight of this water was equal to that of the two airs which had served to form it. (Lavoisier, 1862, pp. 338-339)

At that time, mathematics was regarded as *a priori* knowledge; therefore, Lavoisier's statement "it is no less true in physics than in geometry that the whole is equal to its parts" seems to imply that he also regarded the law of mass conservation as undeniable and *a priori*. Indeed, at one place he explicitly uses the phrase "*a priori*" to describe that idea:

This experiment gave results similar to those of the previous one. Its result was also that when phosphorus burned, it absorbed a little more than one and a half its weight of oxygen, and I have obtained also the certainty that the weight of the new substance that was produced was equal to the sum of the weights of the burned phosphor and of the oxygen that it had absorbed – and that was easy to predict *a priori*. (Lavoisier 1789, vol. 1, p. 63)

In another work, where Lavoisier described the combustion of alcohol, oil and other substances (1784)[11], he presented the law of mass conservation as "evident":

The combustion of olive oil does not contain as much uncertainty as that of the spirit of wine, because olive oil is not capable to volatilize easily; one can know with rigorous accuracy the quantity burned, by difference of the weights determined before and after combustion. [...]
Concerning the water which has been formed, it could neither be collected nor weighed, and I have elsewhere

---

[11] "Mémoire sur la combinaison du principe oxygine avec l'esprit de vin, l'huile et différents corps combustibles" (Lavoisier, 1862, pp. 586-600).

> explained the reason for it; it is the same as for the spirit of wine; I have therefore determined it by calculation, always starting from the supposition that the weight of the materials is the same before and after the operation, which I regard as evident [...] (Lavoisier, 1862, p. 595)

In the specific case of Lavoisier, the law of conservation of weight was neither proved nor derived from experiments, and he did not care to test it. It is clear that he did not regard it as an empirical law nor as something that could be checked. It seemed an *a priori* truth, quite obvious for him.

The reader of this paper might be perplexed: if the equality of the weight of the materials before and after chemical reaction is *evident*, why wasn't it used before Lavoisier?

As a matter of fact, it was much used before the French chemist.

> Most students of chemistry have diligently learned that it was Lavoisier who introduced the concept of the conservation of matter into chemistry, using this principle to dismiss the imaginary element of phlogiston.
>
> Nevertheless, the conservation of matter is a basic assumption that underlies ancient physics. A great majority of Greek philosophers and early modern scientists considered matter to be eternal and indestructible without having any experimental evidence for it. The conservation of matter is so deeply embedded in western science that the philosopher Émile Meyerson considered it as an *a priori* metaphysical assumption and the necessary foundation for all scientific endeavors[12]. (Bensaude-Vincent & Simon, 2008, pp. 115-116)

Once more, it is necessary to remark that many teachers and students of chemistry are misled by the eponym, "Lavoisier's law". The naïve reader might reason thus: "If the law of

---

[12] As will be seen later, this interpretation of Émile Meyerson's views is not correct.

conservation of matter is ascribed to someone called Lavoisier, that person must have been the *first* to prove, or to discover, or to present, or to use it… otherwise it would not have been named thus, isn't it?" No, that is not correct! Beware of eponyms! They are usually meaningless and they convey a misleading view of the history of science (Martins, 2015).

> The concept of conservation of mass was suggested in the section on vinous fermentation. Lavoisier recognized that sugar, which is composed of carbon, hydrogen, and oxygen, is converted by means of yeast to carbon dioxide and spirit of wine, for which he introduced the Arabic term alcohol. He took for granted that it should be possible to account for all the original matter in the final products. In other words, he reduced chemical change to a balance-sheet type of operation when he suggested that the weight of products should be equal to the weight of reactants, an idea utilized but not expressed earlier by Black and Cavendish. Even earlier, the Russian scientist Mikhail Vasilevich Lomonosov (1711-1765) had tacitly assumed that when one body gained weight some other body lost an equivalent amount of weight. (Ihde, 1984, pp. 79-80)

A short account of the relevant contributions of Joseph Black, Henry Cavendish, and Jean Rey, can be found in Robert Whitaker's paper (Whitaker, 1975). Aaron Ihde's account, quoted above, might suggest that other people before Lavoisier had only *implicitly* used the principle of mass conservation, but that they had not clearly or explicitly presented it. That is not true. We can find clear statements of the law before Lavoisier:

> […] we want to insist on the fact that Lavoisier did not invent either the idea of the conservation of mass in a chemical reaction or the use of the scales in chemistry. The idea that nothing can be created or destroyed can be found in the writings of the ancients, and many physicists and chemists had championed it as an axiom before Lavoisier. Van Helmont, for example, explicitly proposed that: "Nothing

comes into being from nothing. The weight comes from another body weighing just as much." (Bensaude-Vincent & Simon, 2008, pp. 85-86)

In quantitative experiments Lavoisier assumed the truth of the law of conservation of matter. He was not responsible for first stating this principle, which goes back to Classical antiquity [...] and was often mentioned before his time. Chardenon in 1764 said: "it is a generally adopted principle, that the absolute weight of a body can be increased only by the addition of new parts of matter. The law of converses therefore points out that it cannot become lighter except by the subtraction of these same parts". (Partington, 1961-1970, vol. 3, p. 377)[13]

Émile Meyerson described one of the oldest uses of arguments similar to those of Lavoisier for finding the weight of something that was not weighed:

[...] one cannot doubt that it [the law of conservation of weight] had been understood and taught in some philosophical schools of Antiquity. A curious passage of a treatise attributed to Lucian [of Samosata] endorses this. "If I burn one thousand pounds of wood, Demonax, how many pounds will there be of smoke?" "Weigh the ashes", he says, "the rest is the smoke that is searched for". Demonax was a Cynic philosopher of the second century of our era. He appears to have been exclusively preoccupied with ethics, theology, and politics. Nothing indicates either that he professed atomistic opinions, nor that he had studied

---

[13] Cited in French by Partington: "c'est un principe généralement adopté, que la pesanteur absolue d'un corps ne peut être augmentée que par l'addition de nouvelles parties de matiere. La loi des contraires indique donc qu'il ne peut devenir plus léger que par la soustraction de ces mêmes parties". Notice that Chardenon's paper containing this statement described his research on the calcination of metals and their changes of weight – a subject that was shortly afterwards studied by Lavoisier.

scientific questions. The quotation is all the more significant, because it proves that this reasoning, analogous to those that we make because of the principle [of conservation of weight], had become current in the philosophical schools. (Meyerson, 1908, p. 137)

One of the few authors of the seventeenth century who stated and used the principle of conservation of weight was Jean Rey. Meyerson conjectured that his thought was influenced by the atomism of the ancient philosophers, since at that time the work of Lucretius (*De Rerum Natura*) had been published and was widely discussed. Rey accepted that all matter has weight, including air; and explained the increase of weight of calcined tin as due to the addition of air (Meyerson, 1908, p. 145).

The history of the law of conservation of matter was the subject of Émile Meyerson's earliest publication: "Jean Rey et la loi de la conservation de la matière" (Meyerson, 1884).

> This law is regarded, after the admirable Lessons on the Philosophy of Chemistry of Mr. Dumas, as Lavoisier's creation. It is undeniable that this was the principle that guided him in all his discoveries and that with its help he reformed Chemistry. However, that law was clearly stated, rigorously formulated and applied to the study of chemical phenomena almost one and a half century before him, by the physician Jean Rey. (Meyerson, 1884, p. 299)

Jean Rey's work was published as a pamphlet with the title "Essays [...] on the search for the cause of the increase of weight of tin and lead when they are calcined", in 1630. In this book, Jean Rey first argues that all the four material elements that were accepted in Antiquity – earth, water, air, and fire – are heavy; that all of them have a tendency to fall; and that there exists no absolute lightness of matter – a tendency to recede from the Earth or from the center of the universe, as taught by Aristotle. Then, Jean Rey presents his principle of conservation of weight:

ESSAY VI. *Heaviness is so closely united to the primary matter of the elements, that when these are changed one into the other they always retain the same weight.*
My chief care hitherto has been to impress on the minds of all the persuasion that air is heavy, inasmuch as from it I propose to derive the increase in weight of tin and lead when they are calcined. But before showing how that comes to pass, I must make this observation that the weight of a thing may be examined in two ways, viz. by the aid of reason, or with the balance. It is reason which has led me to discover weight in all the elements, and it is reason which now leads me to give a flat denial to that erroneous maxim which has been current since the birth of Philosophy—that the elements mutually undergoing change, one into the other, lose or gain weight, according as in changing they become rarefied or condensed. With the arms of reason I boldly enter the lists to combat this error, and to sustain that weight is so closely united to the primary matter of the elements that they can never be deprived of it. The weight with which each portion of matter was endued at the cradle, will be carried by it to the grave. In whatever place, in whatever form, to whatever volume it may be reduced, the same weight always persists. (Rey, 1904, p. 14)

This is a clear and explicit exposition of the principle of conservation of weight.

Jean Rey did not attempt to prove it by experiments, since he regarded it as a product of reason. He tried to demonstrate it by an aprioristic argument, but his reasoning was unconvincing and it will not be reproduced here. In the same work, Jean Rey used the principle to elucidate the increase of weight of calcined tin and lead, explaining it as due to the combination of the metals with air (oxygen was unknown, at that time). Nowadays, his contribution is regarded as a very important scientific contribution. At that time, however, Jean Rey's ideas had no impact.

In the late seventeenth century, Christiaan Huygens stated that weight was the measurement of quantity of matter

(Meyerson, 1908, p. 147); and Isaac Newton introduced the concept of mass as equivalent to quantity of matter. They did not, however, deal with mass (or weight) conservation. Robert Boyle, their contemporaneous chemist, made implicit use of the principle, although he did not express it (Meyerson, 1908, p. 148). In the early eighteenth century, several researchers, including Samuel Cottereau Duclos, Wilhelm Homberg, Pieter van Musschenbroek and George Berkeley, have also made implicit use of the principle (Meyerson, 1908, p. 149). However, belief in the conservation of weight was not consensual and even scientist who made extensive use of the balance, such as Henry Cavendish, denied the principle (Meyerson, 1908, p. 149).

Neither Lavoisier, nor the other scientists or philosophers who accepted the principle of conservation of mass (or weight) before him, attempted to prove it by experiment. They admitted it as an evident or obvious *a priori* truth.

## 7. ÉMILE MEYERSON ON THE LAWS OF CONSERVATION AND THE PRINCIPLE OF CAUSALITY

Although several authors had accepted and used the principle of conservation of weight before Lavoisier, one should not think that *everyone* agreed to it. Although Lavoisier compared it to a mathematical truth (the whole is equal to the sum of its parts), it did not command a general acceptance, as arithmetic or geometry did. To be sure, there were many relevant authors from Antiquity to the eighteenth century who did not accept it – including Aristotle and Descartes (Meyerson, 1908, chapter IV). Were they unable to understand such an obvious truth?

> But generations of scholars and philosophers have advanced opinions clearly implying the negation of the principle. Can we say that the idea never suggested itself to them? It is in itself of great simplicity and can be deduced so directly from the principle of causality that it presents itself, so to speak, invincibly to our mind. (Meyerson, 1908, p. 162)

Émile Meyerson presented for the first time a deep analysis of this historical problem in his book *Identité et Réalité* (1908). His main thesis is that there is a *philosophical* aprioristic principle that was always accepted – the principle of causality – which is ample and vague, impossible to submit to any empirical test because it provides no clear observational prediction. On the other hand, there are several *scientific* principles of conservation that are similar, in form, to the principle of causality; they can be submitted to experimental tests; but they usually seem *a priori* truths because of their resemblance to the principle of causality.

The main subject of Meyerson's book *Identité et Réalité* is the law of causality. When he wrote his work, philosophy of science was strongly influenced by positivistic views that stated that the only aim of science was to *describe* the world, by means of laws; *understanding* the world seemed impossible and/or unnecessary. Meyerson's most general claim was that the search for lawfulness is not sufficient: "science also seeks to *explain* the phenomena, and that explanation lies in the identification of the antecedent and the consequent" (Meyerson, 1908, p. vi). In some sense, everyone agrees that science looks for explanations; however, Meyerson's peculiar understanding of explanation and causality was not the traditional one – it was highly peculiar.

What did Meyerson understand as the principle of causality? He turned his attention to a specific meaning of cause, broadly corresponding to Aristotle's material cause (Drouin 1964) – that is, something constant underlying the phenomena (Follon 1988). Meyerson emphasized the special interpretation of causality that presupposes equality between cause and effect, being equivalent "to the well-known formula of the scholastics, *causa aequat effectum*" (Meyerson 1908, 16). The world is full of changing phenomena; since Antiquity, philosophers (and, later on, scientists) have been searching for something constant behind the mutable events, something that is the same before and after changes occur. That was the motivation of the early

Greek philosophers in the search of the material invisible substratum of nature – and, more specifically, of the atomistic theory (Meyerson, 1908, chapter 2). It is also the motivation behind every law of conservation that has been proposed in science, because they state that something is constant in time, although observable changes are occurring.

> It is from that second principle, the principle of *scientific causality*, that come the atomic theories (chapter II). It also interferes in the laws of science, creating the principles of conservation (chapters III, IV and V) [...] Those conclusions are not a scientific result, they are the outcome of the aprioristic elements that it [science] contains [....] (Meyerson, 1908, p. vi)

According to Meyerson, the search for constancy behind the phenomena is not a result of empirical research – it is a demand of our reason; and it can never be completely fulfilled, otherwise there could be no change in the world.

The principle of causality urges us to search for entities that do not change, in phenomena. The principle of causality is not a description of the world, it is a demand of our reason. In each particular case, we may find – or not – the specific unchangeable entity (or magnitude). However, the world could be built without anything constant or permanent.

> The way followed by our understanding to apply the principle of identity clarifies that it is liable to mistakes, in this matter. There were principles of conservation that have been proposed and that science had to abandon completely, afterwards; or it was necessary to severely transform its content, changing the expression of what is conserved. Along our work we have found examples of both cases. Black's principle of conservation of caloric belongs to the first category: that proposition seems nowadays clearly wrong and, in addition, as contradicted by facts of vulgar experience, such as the heat produced by friction. However, it was for a

long time accepted as securely founded, as one of the most solid grounds of physics. (Meyerson, 1908, p. 370)

When the principle of causality, thus understood, is applied to the concept of matter, it generates the principle of conservation of matter. In its more general form, it states that, although visible matter changes, there is some primordial matter that does not change and is constant throughout all phenomena. The primordial matter of Aristotle was different from the primordial matter of the Greek atomists; but the general principle was the same. During the nineteenth century, some chemists (such as John Dalton) accepted the existence of immutable atoms that remained constant in chemical change; others (such as Wilhelm Ostwald) did not believe that atoms did exist, but nevertheless they admitted chemical unchanging elements – so, the principle of conservation of matter can be satisfied in many different and incompatible forms.

In Antiquity, the Greek and Roman atomists accepted the principle of causality, stating that nothing can be created from nothing, and that nothing can be annihilated. They applied the principle to atoms, which were supposed to be eternal, uncreated, indestructible (Meyerson, 1908, p. 137). Their atoms had weight, and we may suppose that the atomists accepted that weight is conserved, although neither that statement nor its implicit use can be found in their extant writings.

Other Greek philosophers, including Plato and Aristotle, could not accept the conservation of weight. They believed that the four material elements (earth, water, air and fire) can transform in one another; water is heavy and falls to the ground, and it can become air (vapor) that is light is recedes from the ground. In many cases such as this, it was supposed that weight had disappeared or changed. For many philosophers, weight was an accidental quality of matter, as color or shape – and none of those qualities is preserved in the transformations of matter (Meyerson, 1908, p. 139). That way of thinking was dominant during the Middle Ages. The principle of causality was always accepted, and philosophers

> [...] were certainly convinced that something essential in matter, its *substance*, was maintained during its modification. However, since the concept of matter was justifiably separated from weight, it became too difficult to provide a quantitative basis to matter. (Meyerson, 1908, p. 142)

> We find this principle [of conservation of matter] expressed in several quite different ways. Let us discard from the star the formulation "nothing is created, nothing is destroyed" that is still sometimes associated to it and that is obviously too broad: it could be applied also to the conservation of velocity and that of energy – and that is not surprising, because this statement is, as we have seen, just one of the statements of the principle of causality. At least one should say: matter is not created nor destroyed. However, even this statement lacks precision. (Meyerson, 1908, p. 136)

Conservation of matter can be regarded as a *qualitative* principle; or it may be assumed as a *quantitative* principle. In the second case, it becomes the principle of conservation of the quantity of matter. But how should we understand the "quantity of matter" that remains the same before and after the visible transformations of matter? Is it the mass of substances, or some other quantity such as their volume or the number of atoms? The principle of causality is unable to provide any unambiguous interpretation.

A material object has many different properties, such as its size, volume, weight, color, etc. Do all those properties remain the same? Of course not – if *everything* remained unchanged, there could not happen any phenomenon. The principle of conservation of matter cannot state that nothing changes, because that is obviously false. It must indicate that *something* in matter is conserved; it cannot be created or destroyed. Is it the *weight* of matter? Weight is the gravitational force acting upon a body. We know that the weight of an object on the Moon is different from its weight on the Earth; and that its weight on the Earth changes, according to the altitude of the place where it is placed. Therefore, the weight of a material object is not conserved, it may change. We nowadays accept that the *mass* of material bodies is constant. Hence, in a more precise way, the law of conservation of matter is understood as the law of conservation of mass (Meyerson, 1908, p. 137).

The concept of physical mass was only introduced in the seventeenth century. It is not a straightforward concept; and both before and after him, the most common way of describing the principle of conservation of matter is referring to the conservation of weight. Hence, in this paper, we will use "conservation of mass" and "conservation of weight" as equivalent.

> Like everything else we desire mass to be constant; but we desire it more strongly because it is capable of appearing as the essence of matter. (Meyerson 1930, 182)

There is a gap between the principle of causality and any specific law of conservation that cannot be filled by reason alone. The principle of causality is *a priori* and unquestionable. Each particular law of conservation that has ever been proposed in science can be submitted to tests and does not contain the same degree of certainty as the principle of causality.

Let us explain Meyerson's conception of conservation laws in a schematic way. The principle of causality, as understood by him, states: "In every transformation, *something* is permanent or constant." Philosophy can only go as far as that. Science, however, needs more: *what* remains constant in the transformations? How can I detect or measure this "something"? Reason cannot offer the solution. Only empirical research can provide a satisfactory answer. Therefore, every principle of conservation contains two components: the *a priori* structure of the principle, that comes from the principle of causality; and the *a posteriori* interpretation or scientific content, coming from experience.

> Therefore, what is really aprioristic in science is the starting series of postulates that we need for an empirical science, that is, for being able to articulate this proposition: nature is well-ordered, and we can know its route. However, this strictly empirical science is an artificial fabrication and *science* is not rigorously empirical; it is also the application to nature, by successive phases, of the principle of identity, the essence of our understanding. However, from that principle we cannot deduce any precise proposition: and for that reason, a *pure* science cannot exist, counter to what Kant supposed. When trying to explain the phenomena, we attempt to adjust them to what this principle postulates, and for that reason its intervention in science manifests itself as a tendency, the causal penchant. (Meyerson, 1908, p. 369)

## 8. HANS LANDOLT AND THE CONSERVATION OF MASS

Let us return to the principle of conservation of mass. Neither Lavoisier nor his predecessors have attempted to test it. What happened afterwards?

Although Lavoisier did not care to test the principle of conservation of mass, it could have occurred that later researchers did test and confirm it. It seems that most chemists, teachers and students, believe that this did happen. Meyerson, however, denied it:

> What is the current situation regarding this question? Since Lavoisier, the balance has become the preferred instrument of the chemist; and one could say – this is, for instance, Mr. Ostwald's opinion – that in some sense every quantitative analysis undertook by a chemist amounts to a verification of the conservation of matter. However, we must not try to prove too much. Those analyses, in general, agree but *grosso modo*; it is unusual that in any somewhat complex series of operations one does not observe deviations too great to be

> attributed to measuring instruments, as Mr. Ostwald is obliged to admit. (Meyerson, 1908, p. 157)

Most chemists, after Lavoisier, simply accepted the principle of mass conservation, without worrying to submit it to a trial. The situation changed, however, in the late nineteenth century.

> Where an effort has been made to verify directly and with precision the conservation of weight in chemical phenomena, one has not always been successful in obtaining results absolutely confirming this principle. It is well known that anomalies have been noticed quite recently by Mr. Landolt. The results of the German chemist[14], although sometimes questioned, seem to have been received in general with but little skepticism by the scientific world. (Meyerson, 1908, 158)

Hans Landolt's researches on the conservation of mass made a strong impression among chemists and physicists, in the late nineteenth and early twentieth centuries. Nowadays, it is ignored not only by scientists, but also by historians of science. I have presented a detailed description and analysis of this episode in another paper (Martins, 2019b; see also Martins, 1993; Cerruti, 1996). For this reason, it will be briefly presented here.

The motivation for Landolt's experiments was the hypothesis of William Prout (1785-1850) concerning the composition of atoms. According to Prout's proposal, hydrogen should be regarded as the first matter (πρώτη ὕλη) of which the other elements are made (Brock, 1969). Each atom should be composed of an integer number of atoms of hydrogen and, therefore, the atomic mass of any chemical element should be a multiple of the atomic mass of hydrogen.

---

[14] As a matter of fact, Landolt was a Swiss chemist, although he worked in Germany.

The hypothesis was compatible with data available when Prout proposed it, but soon new estimates of the relative atomic weights, by Jons Jacob Berzelius (1779-1848), challenged the conjecture. However, in 1840 Jean Baptiste Dumas (1800-1884) and his student Jean Servais Stas (1813-1891) disputed Berzelius' measurements, providing new evidence that the atomic weight of carbon was almost exactly six times that of hydrogen. However, some atomic weights were clearly fractional, such as those of chlorine and copper. In 1857-1859 Dumas was obliged to introduce a new *ad hoc* hypothesis: that the weight of the fundamental particle was half or one quarter that of the atom of hydrogen (Farrar, 1965, pp. 302-305; Smith,1976, p. 61; Hamerla, 2003, p. 359).

Stas, who initially supported Prout's conjecture, published in 1860 a new determination of eight atomic weights that provided evidence *against* the hypothesis of multiple atomic weights. Jean Charles Galissard de Marignac (1817-1894) immediately defended Prout's hypothesis, claiming that the observed deviations could be due to violations of the law of constant composition, or to the breakdown of the law of conservation of weight in the formation of complex atoms (Farrar, 1965, pp. 306-308; Smith, 1976, pp. 62-63; Hamerla, 2003, pp. 359-360).

Julius Lothar Meyer (1830-1895) claimed that there should be some kind of primordial matter making up the several elements (Farrar 1965, 309; Smith 1976, 54). He conjectured that the atoms of the several elements were built up of hydrogen atoms *together with an ether condensation*. The formation of complex atoms could entail weight variation because of the addition of ether in their synthesis (Smith, 1976, p. 57; Kragh, 1989, pp. 59-60).

Besides Lothar Meyer's conjecture, there is another atomistic theory of the ether of the late nineteenth century that is especially relevant for understanding Landolt's experiments: that of the botanist Carl Wilhelm von Nägeli (1817-1891). In an appendix to his famous book on the theory of evolution, he proposed that the ether was composed of indivisible particles called *Ameren* (in German). According to him, the atoms of each element are built of millions of *Ameren* and, therefore, there is no reason why the law of multiple atomic weights should be valid.

Hans Landolt began his experiments on weight conservation in chemical reactions in 1890, motivated by the discussion on Prout's hypothesis. In his first paper on this subject (Landolt, 1893) he referred to Prout's hypothesis, to Marignac's conjectures, to Lothar Meyer's speculations and to Nägeli's ether theory. The rationale of his experiments can be reconstructed thus: *if* there is any change of weight related to the production of the atoms of the several elements, *then* there could also exist some change of weight in chemical reactions that do not involve the production of atoms, because of the possible change of ether condensations in the formation or decomposition of compounds.

In his first experiments, Landolt studied four different chemical reactions:

$$Ag_2SO_4 + 2FeSO_4 = 2Ag + Fe_2(SO_4)_3$$

$$KIO_3 + 6H_2SO_4 + 5KI = 3I_2 + 6KHSO_4 + 3H_2O$$

$$I_2 + 2Na_2S_2O_3 = 2NaI + Na_2S_4O_6$$

$$Cl_3C\text{-}CH(OH)_2 + KOH = Cl_3CH + HCO_2K + H_2O$$

The chemical reagents were enclosed in special hermetically sealed glass recipients. For each experiment, Landolt used two similar systems, with the same set of reagents. Their weights were closely similar, with a difference of just a few milligrams – for a total weight of about one kilogram.

To understand his procedure, let us call the two systems A and B. Before the chemical reaction had occurred in any of them, the weight difference was measured several times, in the course of a few days. Then, Landolt produced the chemical reaction in A. After the reaction was complete, he compared again the weights of A and B. Then, system B was manipulated and the chemical reaction occurred in the second device. After the reaction was complete, the weights of A and B were compared again. Landolt made a detailed study of systematic influences, taking into account changes of the moisture at the surface of the glass flasks, their changes of temperature, etc.

In 1890 he studied the first reaction (reduction of silver sulphate). In each set of measurements, the standard deviation was of the order of 0.01 mg. After the chemical reaction had occurred in the apparatus A, its weight was observed to be *smaller* than before, with a reduction of 0.167 mg. After the chemical reaction in B, its weight also decreased, by 0.131 mg. The changes were much larger than the random errors, and therefore they seemed significant. A repetition of this experiment produced similar results. Therefore, in the case of the first reaction, there was an apparent reduction of about 1/800,000 of the weight of the chemical substances, or 1/6,000,000 of the weight of the devices. For the other three reactions he studied, the observed changes were smaller and somewhat irregular.

Landolt was a highly respected chemist and his research seemed careful enough to deserve attention. His article was soon reproduced in two other scientific journals. There was immediate reaction –papers presenting new experiments with positive or negative results, and theoretical discussions.

In 1906 Landolt published his second paper on the apparent changes of weight in chemical reactions. Prompted by criticism and suggestions made by several authors, he obtained a new balance, introduced changes in the apparatus and took new precautions. In the new series of experiments, Landolt also observed significant reductions of weight in the reduction of silver sulfate or nitrate by iron sulfate. The mean change was a weight reduction of 0.29 mg for each 100 g of silver, and the change seemed proportional to the total weight of the chemicals.

In the case of other reactions, the observed effects were smaller and sometimes no change was noticed. He commented:

> The question now is how the weight loss can be explained. First of all, one can express the suspicion that there is still an external cause which has not yet been discovered, but taking into account the care with which all possible sources of error have been investigated, this view is unlikely to have any probable value. On the other hand, the fact that the change only occurs to a marked extent in certain reactions, such as the reduction of silver or iodine, and in others is slight or absent, strongly suggests a relation to the chemical process.
>
> Since the explanation must be such to account only for weight loss and never increase, no other hypothesis remains except the one already mentioned in the introduction, according to which the phenomenon is founded on the detachment of small mass particles from the chemical atoms. (Landolt, 1906, p. 619)

In principle, Landolt's hypothesis in 1892 was similar to that of 1906: in both articles, he supposed that some very small subatomic particles had escaped from the glass vessels. However, in his first paper he was thinking about *ether particles*, and in the second one he referred to the recently discovered *radioactive phenomena*. He conjectured that a small part of the atoms could split off, as the result of release of energy during chemical reactions. To account for the observed effect, the emitted particles should traverse the walls of the glass flasks. Therefore, the relevant problem was whether the glass vessels could be regarded as closed systems of not, regarding ponderable substances.

Notice that Landolt always accepted the strict conservation of weight in chemical reactions. He did state that the weight of the apparatus had changed, but he interpreted the change as due to something that had escaped from it – ether particles or subatomic particles. However, if those auxiliary hypotheses encountered severe difficulties and if there was no other alternative explanation for the change of weight during chemical reactions, then the experiment itself could have been faulty. That was Landolt's attitude in his last works on this subject, published in 1908 and 1910.

Although in his previous researches Landolt had paid much attention to all possible sources of systematic error, in his 1908 paper he studies again two of them: the variation of moisture on the outer walls of the glass vessels, due to the heat developed by the chemical reactions; and the change of volume of those containers. He concluded that the first effect did not introduce any relevant error; however, the second one did, because the thermal dilatation of the glass vessels was irreversible, contrary to his expectations. The increased volume produced an apparent decrease of the weight of the system, because it was weighed in the air – not in a vacuum.

Because of this possible source of error, Landolt repeated his experiments and he also computed corrections that should be applied to his former measurements. His final conclusion, taking into account 48 experiments concerning 15 different chemical reactions, was that there was no weight change larger than the estimated experimental error of 0.03 mg for a total weight of 400 g. If there was any real change of weight, it was smaller than one part in 10 million.

Landolt's last work on this subject was published posthumously (Landolt, 1910). It was a review of all his previous experiments, with a conclusion similar to that of the 1908 paper. Some later authors described Landolt's research as a *proof* of the law of conservation of matter (or weight) in chemical reactions (Laue, 1959, pp. 509-510). Of course, from a logical point of view, it is impossible to provide an experimental proof that the conservation of weight holds *exactly*, for *any* chemical reaction. All that can be safely admitted is that *for a specific set of chemical reactions*, studied under such and such conditions, no weight change *larger than the experimental error* was observed.

## 9.  MEYERSON'S INTERPRETATION OF THE PRINCIPLE OF MASS CONSERVATION

After presenting an outline of the history of the principle of mass conservation, Meyerson discusses whether it is an empirical law (Meyerson, 1908, pp. 154-158) or an *a priori* principle (Meyerson, 1908, pp. 158-162), providing strong arguments against both alternatives; and then he explains his own view (Meyerson, 1908, pp. 162-168).

> Is this principle of empirical origin? This has often been affirmed, and John Stuart Mill especially formulated this thesis with much clearness. According to him, the conservation of matter is suggested to us from the very beginning of our observations, so to speak, by a large number of concordant phenomena, whereas others, on the contrary, seem to contradict it. The hypothesis was formulated that this principle was, not partially, but entirely, true, and it was verified afterwards. The verification having succeeded, the principle was established, exactly as any other experimental law. (Meyerson, 1908, p. 154)

This is, of course, the standard empiricist view of the founding of scientific laws. The principle of conservation of mass would have been obtained by experiments, that is, *a posteriori*. Historical information, however, disproves this interpretation. From the Greek atomists to Lavoisier, no researcher attempted to test the law.

> [...] it is enough to peruse his [Lavoisier's] works to be convinced that this is not so; that, exactly as the ancients and as Jean Rey, he applies the principle with complete confidence, not doubting for a single moment that it should be confirmed by experience, and that every anomaly must be only apparent and should be explicable. (Meyerson, 1908, p. 156)

It follows from the information that we have just briefly summarized that even at the present moment the certainty with which the principle of the conservation of matter appears invested is much higher than what is permitted by the experiments which are supposed to serve as its basis. (Meyerson, 1908, p. 158)

After Immanuel Kant's philosophical work, it had become common to distinguish scientific statements as *a priori* (that is, evident and produced by reason) or *a posteriori* (that is, not evident, not purely rational and depending on experience). Having shown that the principle of conservation of matter was not grounded on experiments, Meyerson discussed the second possibility:

Is this principle therefore aprioristic? This seems to be the most widely diffused opinion in our time among scientists and philosophers, an opinion sometimes clearly stated, sometimes implicitly affirmed. Moreover, [...] we shall not have to search long for the source of this opinion; all those who have endeavored to demonstrate *a priori* the conservation of matter have brought directly into play the principle of causality, to such a point that, as we have seen at the beginning of the chapter, the very statements of the two principles are sometimes muddled up. (Meyerson, 1908, pp. 158-159)

Kant bothered himself with this question many times. [...] He explained himself in a more detailed way in the *First Metaphysical Principles of Science and Nature*. There he expresses his "first theorem of mechanics" thus: "Through all the modifications of material nature, the total quantity of matter remains the same, without increase and without diminution." In the "demonstration" of this theorem, Kant makes use of the principle that he borrows from "general metaphysics," which states that in all the modifications of nature, there is neither creation nor destruction of substance. Afterwards he determines that substance, for matter, is its quantity [...] (Meyerson, 1908, p. 159)

In the specific case of Kant, Meyerson shows that there was a gradual transformation of the general principle of causality to the specific statement of conservation of mass or weight. However, that transition requires some steps that cannot be justified by pure reason: the identification of "substance" with "quantity of matter" and then with weight of mass is unjustifiable *a priori*. Meyerson also discussed the *a priori* "proofs" of other authors (Arthur Schopenhauer, William Whewell and Herbert Spencer) and showed that they also mistook the empirical quantitative concepts of weight or mass for the philosophical concept of substance (Meyerson, 1908, pp. 160-161). Therefore, those aprioristic "proofs" are unacceptable. Meyerson also pointed out that many philosophers had clearly denied the principle; and that the circulation of atomistic ideas and, later, of Jean Rey's work did not lead to the acceptance of the principle of conservation of matter (Meyerson, 1908, p. 162). If that were really an aprioristic principle, why wasn't it accepted by everyone?

> The mystery disappears if we think what really is the aprioristic demonstration of the principle. It is completely based on causality. Now, what we call causality is only a tendency, the tendency to retain the identity of some things in time. At most one may say that the causal tendency makes us the hope that these things are such that we may, without too great a violence, regard them as essential. [...]
>
> At the basis of the principle of conservation of matter there are three distinct notions: matter, weight, and mass. Matter is a common-sense notion, a complex one, which synthetizes an infinite number of properties. It is clearly contrary even to the most superficial experience to suppose the conservation of all those properties. Therefore, in sustaining the conservation of matter, one only postulates the permanence of some, among them; that is the reason why this statement cannot interest science as long as one does not elucidate precisely what must be conserved. (Meyerson, 1908, p. 162)

It is possible to measure many different properties of matter: its temperature, its size, its thermal capacity, etc. Which one is conserved, and why? This can only be found *a posteriori*, by experiments. There is no logical connection between mass (or weight) and the essential notion of matter. However, when one overlooks that gap, it seems that it is possible to derive the law of conservation of mass (or weight) from the *a priori* principle of causality and that no empirical assessment is needed.

> [...] any attempt to deduce *a priori* the principle of the conservation of weight will certainly be sterile. The opposite illusion can only have its source in the causal tendency; and it is this same tendency, which in ancient atomists, in Jean Rey, probably in Lavoisier himself, and surely in his contemporaries, has contributed to the rise of the principle, permitting its formulation, without proof, as a self-evident truth, ensuring its dominance.
>
> We desire mass to be constant, like everything else; but we desire it more strongly because it appears to us as the essence of matter. (Meyerson, 1908, p. 162)

> [..] the conservation of mass cannot be considered as an aprioristic truth; and we believe that the assembly of the considerations which we have brought out clearly predisposes to establish that it is a plausible proposition, just like the principle of inertia. (Meyerson, 1908, p. 163)

Meyerson's approach partially agrees with that of Kant:

> The true approach was pointed out by Kant: there is indeed a concord between our understanding and reality, but this agreement is partial [...] Reality is partly intelligible and our scientific knowledge contains a mixture of aprioristic elements and other that are *a posteriori*. (Meyerson, 1908, p. 366)

However, Meyerson criticized Kant for assigning an excessive weight to deduction:

So, when he talks about conservation of matter, he says: "We borrow from general metaphysics this principle that we accept as basic: that in every change of nature, no substance is lost or created. Here one is just displaying what is substance, in matter." That is, indeed, as we have seen, the real foundation of that principle. However, for Kant, the last part of its propositions is likewise aprioristic: the concept of matter includes, in his opinion, not only that of mass, but also that of weight [...]

Kant believes that science contains a *pure* component, that is, a purely rational one and, consequently, completely *a priori*. That part contains not only what we call cinematics [...] but also part of mechanics. However, that is not the case. There is no pure mechanics, nor pure cinematics. (Meyerson, 1908, pp. 366-367)

There are scientific laws that can be submitted to experiments but that, nevertheless, are believed to be universal truths, because they seem *necessary*. Meyerson accepted William Whewell's interpretation of this apparent contradiction:

"The solution of that paradox is this: those laws are interpretations of the axioms of causality. The axioms are universally and necessarily true, but the appropriate interpretation of the expressions they contain is taught by experience. Our idea of cause provides the *form* and experience the *matter* of those laws." (Whewell, *apud* Meyerson, 1908, pp. 367-368)

According to Meyerson, the scientific laws of conservation are neither *a priori* nor *a posteriori*, although they contain components of the two types. In the lack of a better terminology, he described them as *plausible* laws.

[...] taken literally, the principle of identity in time would signify: everything persists, an affirmation immediately denied by experience; [...]. Hence the statement becomes: certain essential things persist. But this is an indefinite formula, for it does not tell us what are the things which persist and which, consequently, we ought to consider as essential. It is experience alone which can teach us that. But in this matter experience plays a peculiar role, in the sense that it is not free, for it obeys the principle of causality, which we may call more precisely the causal tendency because it manifests its action in commanding us to seek in the diversity of phenomena something which persists. The formula constitutes thus, according to Boutroux's admirable expression, not a law but a "template" of laws.

From what we have just contended [...], we can draw this general conclusion: every proposition stipulating identity in time appears to us as invested *a priori* with a high degree of probability. It finds our minds prepared, it seduces them, and is immediately adopted, unless contradicted by very evident facts. Perhaps it would be wise to apply to statements of this category, intermediary between the *a priori* and the *a posteriori* a special term. We should propose, for lack of a better one, the term *plausible*. Therefore, every proposition stipulating identity in time, every law of conservation, is plausible. (Meyerson 1908, 133-134)

Meyerson used "plausible" for describing those statements that are neither entirely *a priori* nor totally *a posteriori*. That was not a wise choice, in my opinion, because "plausible" is usually understood as equivalent to probable. With the exception of this name, I agree with Meyerson's interpretation of the laws of conservation.

## 10. WHO WAS ÉMILE MEYERSON?

The interpretation used in this paper was heavily influenced by the work of Émile Meyerson. Since this philosopher is not well known nowadays, this section will present some biographical information about him.

Emil Azriel Meyerson (Fig. 6), better known by the French variant of his name (Émile Meyerson), was born in Lublin,

Poland (then a Russian territory), in 1859, from Jew ancestors.[15] In 1870 he moved to Germany to complete his education, and he later attended the University of Heidelberg, where he studied chemistry with Robert Wilhelm Eberhard von Bunsen (1811-1899) and Hermann Franz Moritz Kopp (1817-1892), and afterwards with Carl Theodore Liebermann (1842-1914), at the University of Berlin (Telkes-Klein, 2007a).

**Fig. 6** Émile Meyerson

Although his early involvement with chemistry was mostly scientific and technical, he soon became interested in the history

---

[15] Émile Meyerson's life and the evolution of his thought have been recently elucidated thanks to the study of his correspondence and manuscripts kept at the Central Zionist Archives in Jerusalem. See, for instance, Bensaude-Vincent & Telkes-Klein (2016).

of that science. During his stay in Heidelberg, he came across the books on history of chemistry written by Kopp, and he was strongly influenced by them. From the history of science, he soon became involved in philosophy of science. His chemistry background clearly inspired much of his later work, as was also the case with other French historians and philosophers of science (Bensaude-Vincent, 2005; Brenner & Henn, 2013).

In 1882 Meyerson moved to Paris, where he was accepted as trainee of Paul Schützenberger (1829-1897) at the *Collège de France*, thanks to a recommendation letter written by Bunsen. Schützenberger was the professor of mineral chemistry at the *Collège de France* and the first director of the *École Supérieure de Physique et de Chimie Industrielles*, where Pierre Curie worked and Marie Curie later began her first researches on radioactivity. With Schützenberger, Meyerson acquired an expertise in industrial chemistry and the synthesis of dyes – one of the most profitable fields of chemistry, at that time. Parallel to his chemical studies, after his arrival to Paris, Meyerson frequently visited libraries to dedicate himself to the study of the history of chemistry. At this time, he became aware of the French version of the history of chemistry, sponsored by Jean-Baptiste Dumas (1800-1884) in his *Leçons sur la philosophie chimique*, which presented Lavoisier as the first chemist in history. He recalled Kopp's works that had shown the importance of pre-Lavoisier phlogiston theory, and began to study the role of some of the precursors of modern chemistry.

At the beginning of 1884, Meyerson visited Lublin and, during his stay there, he wrote his first article, a historical essay on Jean Rey and the conservation of matter (Meyerson, 1884).

Returning to France, he was hired by the chemical industry *Établissements A. Collineau et Cie*, in Argenteuil, where he worked on the production of dyes. After two years he was appointed as a chief chemist of the industry, with double salary; but in July 1886 he resigned this job. In January 1888 he registered a patent of a method for the synthetic production indigo, but the process was not successful on an industrial scale (Telkes-Klein, 2007a; Bensaude-Vincent, 2010a).

During the subsequent years he maintained some work on chemistry, both in France (analysis of wine and milk) and in Russia (the plan for a chemical industry in Saint Petersburg). Parallel to that work, he kept his scholar inquiry and published a few papers on the history of chemistry (Meyerson, 1888; Meyerson, 1891). Thanks to his knowledge of many languages (Polish, German, Russian, French, English, and Italian), in 1889 he became the foreign policy editor at the *Agence Havas*, one of the oldest news agency of the world. This job did not require a great effort from him – just a few hours a day – and gave him some financial safety in the next nine years, during which interval he dedicated himself deeply to his historical and philosophical research. He spent most of his free time at the *Bibliothèque Nationale*, reading and taking notes about philosophy, science, and the history of science. He was already working on his main philosophical project that would finally produce his masterpiece *Identité et réalité*, but he felt many difficulties and postponed it for years. During this period, besides reading about philosophy, he studied Latin (a language he did not know well) and he got a firm knowledge of mathematical physics. Meyerson's erudition was to become legendary (Telkes-Klein, 2010).

In the decade of 1890 Émile Meyerson got gradually involved in Zionism (Telkes-Klein, 2015). In 1893 he became a member of the organization *'Hovevei Sion* (Lovers of Zion), which had the aim to promote Jewish immigration to Palestine, and advance Jewish settlement there. In 1894 he was the secretary of the *Conférence des Sociétés Palestiniennes*, and from 1894 to 1896 the secretary of the Central Committee of the *'Hovevei Sion* in Paris. In 1898 Meyerson got a job at Baron Maurice de Hirsch's *Jewish Colonization Association* that aimed at the creation of Jew colonies in several parts of the world. He also got involved with Baron Edmond Benjamin James de Rothschild's personal endeavor of establishing Jew colonies in Palestine. His job required him to travel to Russia, Poland and Palestine. He also published several works concerning the situation of Jews in Russia, and about the colonies established in Palestine. Meyerson became the director of one of its divisions, the *Jewish Colonization Association for Europe and Asia.*

A letter from Émile Meyerson to his sister shows that his decision to work at the *Jewish Colonization Association* was a financial choice (Telkes-Klein, 2010; Telkes-Klein, 2004, p. 206). His earnings at *Agence Havas* were low and he had debts. He was nearly 40 years old and had no hope of obtaining an academic position. He needed a more comfortable situation, although he knew in advance that he would not have so much free time in that new position as he had during the period of *Agence Havas*.

He never became a university professor. He was not, however, a secluded scholar. He involved himself in the *Société Française de Philosophie*, founded by Xavier Léon in 1901, and maintained a circle of friends with whom he kept a continuous discussion of philosophical ideas (Soulié, 2010). One of them was Lucien Lévy-Bruhl, who is nowadays better known for his anthropological and sociological studies, but who was at that time the professor of history of modern philosophy at the Sorbonne.

The first and most influential of Émile Meyerson's philosophical works was *Identité et réalité*, published in 1908 at the expenses of the author. The book received favorable reviews in several French and foreign journals. It established Meyerson as an outstanding philosopher and it immediately called the attention of Léon Brunschvicg, Henri Bergson, and Ernst Cassirer, to cite only a few remarkable names.

Meyerson became very prominent, not only through his publications but also because of an intellectual circle that gathered around him, including philosophers and scientists. After his retirement from the *Jewish Colonization Association*, in 1923, Meyerson promoted a weekly meeting at his home, every Thursday afternoon, regularly attended by young and older scholars, including Louis de Broglie, Léon Brunschvicg, Alexandre Koyré, André Lalande, Paul Langevin, Lucien Lévy-Bruhl, André George, Salomon Reinach, Henri Gouhier, André Metz, and Hélène Metzger (Telkes-Klein, 2007a; Telkes-Klein, 2007b).

Some of the participants of this circle can be called his disciples or *protégés*: Henri Gouhier, Alexandre Koyré, André Metz, Hélène Metzger, Henri Poirier, Désiré Roustan. Hélène Metzger was a niece of Lucien Lévy-Bruhl. She became a remarkable historian of chemistry and dedicated one of her books, *Newton, Stahl, Boerhaave et la doctrine chimique* (1930) to Meyerson (see, however, Chimisso & Freudenthal, 2003). Koyré referred to Meyerson as his teacher, and his first outstanding book on the history of science, *Études Galiléennes*, published in 1939, was dedicated to the memory of Émile Meyerson (Simons, 2017).

Besides the three editions of his first book, *Identité et réalité*, Meyerson published many articles and several books during his lifetime: *De l'explication dans les sciences*, 2 vols. (1921), *La déduction relativiste* (1925), and *Du cheminement de la pensée*, 3 vols. (1931). *La déduction relativiste* was written as a reaction to Albert Einstein's theory of general relativity, offering a philosophical interpretation of the role of space-time in that theory. In 1928, Einstein published a joint article with André Metz where he expressed approval and admiration for that work (Einstein & Metz, 1928). In 1930, Meyerson's *Identité et Réalité* had been translated to German, English, and Spanish, becoming available to many readers.

Shortly before his death, Meyerson wrote his last work, *Réel et déterminisme dans la physique quantique* (published in 1933), by request of his friend Louis de Broglie, who wrote the preface of the book. A posthumous volume, *Essais*, edited by Lucien Lévy-Bruhl in 1936, contains a selection of his articles.

Émile Meyerson's work was later strongly criticized by Gaston Bachelard (Lecourt, 2002, pp. 28, 35-39; Perraudin,

2008; Wetshingolo, 1996), maybe unfairly (Fruteau de Laclos, 2008; Bensaude-Vincent, 2010b), and this was possibly one reason why he was overlooked in the following decades in France. In other countries, the influence of Analytic Philosophy was certainly a motive for neglecting Meyerson's ideas (Biagoli, 1988, pp. 35-37). His importance was, however, vindicated by Thomas Kuhn, who referred to Meyerson's ideas as influential while developing the ideas for his main work *The Structure of Scientific Revolutions* (Kuhn, 2012, p. xl). In the recent decades, Meyerson's philosophy has received much more attention (Laugier, 2009; Brenner, 2010). Some of his letters and unpublished manuscripts have been recently published by Bernadette Bensaude-Vincent and Eva Telkes-Klein (Meyerson, 2009; Meyerson, 2011), and his books have received new translations.[16]

## 11. THE NATURE OF THE LAW OF MASS CONSERVATION AND SCIENTIFIC EDUCATION

The historical study of the law of conservation of mass provides a noteworthy instance of the influence of philosophical ideas in science. The philosophical principle of permanence of substance – Émile Meyerson's law of causality – is not a result of science; it is an *a priori* principle that has shaped the development of the laws of conservation.

This case study is relevant for the teaching of chemistry (and of science, in general), because it clearly shows, in a particular historical episode, the influence of philosophical principles on the development of science, thereby providing a nice example against the inductivist view of science.

There are, however, many difficulties in any attempt to convey this epistemological message in science teaching. Both teachers and students have their preconceptions about what is

---

[16] There is a fairly complete bibliography on Meyerson, describing works published up to the beginning of the current century (Fruteau de Laclos, 2003).

science and about how scientists work, and most of them accept naïve inductivism. It is very hard to replace those preconceptions by a more adequate understanding of science – teachers and students usually react against any deep change of their views, or they simply disregard the new message and the arguments that are presented for it (in this case, historical information).

Any long-term educational change must involve both new educational materials and techniques, and the adequate training of teachers. The second point is probably the hardest challenge. University professors who instruct prospective science teachers usually have the same inadequate knowledge of history and philosophy of science that should be supplanted. All over the world, the number of historians and philosophers of science is negligible compared to that of scientists; their direct influence at the universities upon potential science teachers cannot be considerable. It would be possible, however, to strengthen their influence if they provided an adequate training to the future university professors who will instruct the potential new teachers.

Although digital media are increasingly used as educational materials, textbooks have not been superseded. Unfortunately, scientific textbooks show a strong apathy to the introduction of an adequate view on the nature of science, and they tend to perpetuate the old-fashioned views regretted by us – those who endorse the use of history and philosophy of science in education.

There are some outstanding exceptions. In a well-known textbook written by two historians of physics[17] – Gerald Holton and Stephen Brush – we find a detailed and nice account of the history of the law of conservation of mass, including the

---

[17] Holton and Brush's book was the third version of Holton's *Introduction to Concepts and Theories in Physical Science*, where we find a smaller version of the history, which does not mention Landolt (Holton, 1952, pp. 279-285).

contribution of Hans Landolt (Holton & Brush, 2001, chapter 15, pp. 203-208). Unfortunately, their careful historical approach is uncommon.

Historians have sometimes criticized the pseudo-historical lore appearing in textbooks, with little impact. Let us describe two examples concerning the specific case of the principle of conservation of mass.

Bernadette Bensaude-Vincent and Nicolas Journet, in a paper describing the contributions of Lavoisier, mentioned five French textbooks and pointed out some of their historical mistakes, such as ascribing to Lavoisier the "discovery" of the principle as a consequence of "delicate measurements"; the association between the principle and the atomic model (Lavoisier was not an atomist); and the supposed quantitative confirmation of the principle in Lavoisier's experiments on the composition of air (Bensaude-Vincent & Journet, 1993, p. 62).

At the end of his paper on the history of the principle of mass conservation, the historian of science Robert Siegfried commented that "The history of science can teach fundamental lessons about the nature of scientific thought itself" (Siegfried, 1989, p. 22), and then presented the flawed account he found in Ebbing's popular textbook on General Chemistry.

> Antoine Lavoisier (1743-1794), a French chemist, insisted on the use of the balance in chemical research. His experiments demonstrated the law of conservation of mass, *a principle that states that mass remains constant during a chemical change (chemical reaction)*. A flash bulb gives a convenient illustration of this law. (Ebbing, 1987, p. 3, *apud* Siegfried, 1989, p. 23)

Siegfried commented:

> The author completes his point by indicating that the flash bulb weighs the same before and after it is ignited. But what will the student learn from this passage? First, that Lavoisier demonstrated the law of conservation of mass or weight,

presumably in a manner like that utilized in the flashbulb experiment. As we have seen, Lavoisier did no such thing, but took the principle as "an incontestable axiom" incapable of direct experimental demonstration.

But the important point here is not the author's misrepresentation of Lavoisier's work (though that is lamentable enough), but that in so doing he misrepresents the manner by which such broad general principles are established in science. By implying that Lavoisier arrived at this principle by generalization from a large number of cases, presumably some 18th century equivalent of flash bulbs, the author is promoting the Baconian or inductive method, a view long recognized as inadequate and misleading. (Siegfried, 1989, p. 23)

Unfortunately, two decades after the publication of Siegfried's critical comment, the new editions of Ebbing's book still contain the same epistemological mistake:

Antoine Lavoisier (1743-1794), a French chemist, was one of the first to insist on the use of the balance in chemical research. By weighing substances before and after chemical change, he demonstrated the *law of conservation of mass*, which states that *the total mass remains constant during a chemical change (chemical reaction)*. (Ebbing & Gammon, 2017, p. 6)

Unfortunately, misconceptions about the nature of science die hard.

Suppose that some teachers or students are trying to learn about the history of the principle of mass conservation. They would probably start by reading the versions contained in textbooks, popular works on history of science (especially biographical romances of "great scientists") and information available at the Internet. The historical and epistemological facets of those accounts will be probably flawed; but the readers will be usually unable to notice it. Of course, there are nice historical papers on the subject; however, as Gerald Holton remarked,

> Among the historians of science, of which there are only a few thousand professionals in the world, the writings in their professional journals are almost by definition of the kind that rarely would find their way into the hands of science educators. (Holton, 2003, p. 603)

It is frustrating to notice that academic work on history of science, intended to be used by educators, do seldom get noticed and used. Many years ago, I coauthored a paper, in Portuguese, on Lavoisier and the conservation of mass (Martins & Martins, 1993). The full paper is freely available at the Internet, from three different websites. Unfortunately, a fresh Google search provided only 37 web pages referring to that paper – including those pages that contain the paper itself. A slightly worse result was obtained searching for my paper on Hans Landolt's researchers on mass conservation (Martins, 1993).

Maybe the time has come when historians and philosophers of science who intend to produce any influence upon science teaching should devote themselves to the development of more effective strategies to disseminate a better view of the nature of science. It is not enough to produce adequate accounts: it is indispensable that they reach the students and teachers. A promising move would be contributing new information and correcting mistakes appearing in one of the most popular resources of the Internet: the Wikipedia.

In 2008, a graduate student called Sage Ross published an article where he called the attention of historians of science to the relevance of Wikipedia (Ross, 2008). As a rejoinder to this article, the historian of science Alan Rocke summoned up the members of the History of Science Society to review Wikipedia articles on historical material containing manifest blunders (Rocke, 2008).

What is the significance of Wikipedia's articles? Wikipedia is one of the ten top sites, worldwide. Links to Wikipedia pages usually appear at the top of Google searches, when one looks for scientific information. Updating Ross' data, we can remark that the English version of the article on Albert Einstein was retrieved about nine million times in 2017, a mean of about 26,000 times each day – therefore, it has a huge impact. The article on Lavoisier is not so popular, but there were about 450,000 page views in 2017, for the English version. Hence, a mean of over one thousand people find faulty historical information on Lavoisier, daily, when they consult the English version of Wikipedia.

Creating an independent web page on Lavoisier or publishing a paper on his contributions (such as the present paper) will have a much smaller impact than correcting the corresponding article on Wikipedia. Anyone can contribute to Wikipedia, either creating new articles or introducing changes in the available pages. If the modifications are accompanied by relevant scholarly bibliographic references, they will be accepted and preserved, as a rule. By allocating a few hours to improve any popular Wikipedia article such as that, a person can instantaneously reach a large readership and may help to improve the public understanding of the nature of science.

Besides the proposals and advices of Ross and Rocke on the subject, I would like to add another recommendation. I strongly suggest that any historian of science who is willing to contribute to Wikipedia should adopt the strategy of keeping (and criticizing) faulty views, instead of simply replacing them by better interpretations. Indeed, pseudohistory will not disappear only by presenting a sound historical version of the same episodes. Readers must be told about those faulty historical and epistemological versions, and they should also be informed about the reason why these should be rejected. After that, the more acceptable view should be introduced, also explaining why it is better than the other ones.

In my teaching practice, I have adopted a similar approach, and I can tell you that it does work (Martins, 2018). Of course, this is just one among many ways of improving the public understanding of the nature of science.

## 12. FINAL REMARKS

The aim of this paper was not to *defend* Émile Meyerson's interpretation of the laws of conservation (and, particularly, of the principle of conservation of mass). Meyerson's ideas do not need my support – they were well justified by their author. Unfortunately, Meyerson's prolix style did not help in the diffusion of his thought; and his work is not well known nowadays, either in France or abroad. I believe that his peculiar epistemological ideas, grounded upon sound historical research, deserve to be disseminated – and that they can help to prevent scientists, teachers and students from believing in the naïve empiricist interpretation of science.

## ACKNOWLEDGEMENTS

The author is grateful for support received from the Brazilian National Council of Scientific and Technological Development (CNPq).

## BIBLIOGRAPHIC REFERENCES

AYKROYD, Wallace Ruddell. *Three philosophers: Lavoisier, Priestley and Cavendish*. London: Heinemann, 1935.

BARNETT, Martin K. The development of the concept of heat. *The Scientific Monthly*, **62** (2): 165-172; **62** (3): 247-257, 1946.

BENSAUDE-VINCENT, Bernadette; JOURNET, Nicolas. Rien ne se perd, rien ne se crée, tout se pèse. *Les Cahiers de Science et Vie*, (14): 42-62, Avril 1993.

BENSAUDE-VINCENT, Bernadette. Émile Meyerson chimiste philosophe. Pp. 67-90, in TELKES-KLEIN, Eva; YAKIRA, Elhanan (eds.). *L'histoire et la philosophie des sciences à la lumière de l'oeuvre d'Émile Meyerson (1859-1933)*. Paris: Honoré Champion, 2010 (a).

BENSAUDE-VINCENT, Bernadette. Meyerson rationaliste? *Corpus: Revue de Philosophie*, **58**: 255-274, 2010 (b).

BENSAUDE-VINCENT, Bernadette; SIMON, Jonathan. *Chemistry, the Impure Science*. London: Imperial College Press, 2008.

BENSAUDE-VINCENT, Bernadette; TELKES-KLEIN, Eva. *Les identités multiples d'Émile Meyerson*. Paris: Honoré Champion, 2016.

BEST, Nicholas W. Lavoisier's "Reflections on Phlogiston". *Foundations of Chemistry*, **17** (2): 137-151, 2015; **18** (1): 3-13, 2016.

BIAGIOLI, Mario. Meyerson: science and the "irrational". *Studies in History and Philosophy of Science*, **19** (1): 5-42, 1988.

BOYER, Carl B. History of the measurement of heat I. Thermometry and calorimetry. *The Scientific Monthly*, **57** (5): 442-452, 1943.

BRENNER, Anastasios. Sur Émile Meyerson et son œuvre. *Revue Critique. Revue d'Histoire des Sciences*, **63** (1): 299-302, 2010.

BRENNER, Anastasios; HENN, François. Chemistry and French philosophy of science. A comparison of historical and contemporary views. Pp. 387-398, in ANDERSEN, Anne et al. (eds.). *New challenges to philosophy of science*. Dordrecht: Springer, 2013.

BROCK, William Hodson. Studies in the history of Prout's hypotheses. *Annals of Science*, **25** (1): 49-80; (2): 127-137, 1969.

BUNGE, Mario. *Scientific Research II: The Search for Truth*. Berlin: Springer, 1967.

CERRUTI, Luigi. Atomi, elementi chimici, etere ponderabile, modelli ed esperimenti di fine ottocento. In TUCCI, Pasquale (ed.). *Atti del XVI Congresso Nazionale di Storia della Fisica e dell'Astronomia*. Como: Centro di Cultura Scientifica Alessandro Volta. Available online:

<http://www.sisfa.org/pubblicazioni/atti-del-xvi-convegno-sisfa-como-1996/> 1996.

CHIMISSO, Cristina ; FREUDENTHAL, Gad. A mind of her own. Hélène Metzger to Émile Meyerson, 1933. *Isis*, **94**: 477-491, 2003.

EBBING, Darrell D.; GAMMON, Steven D. *General Chemistry*. 11th edition. Boston: Cengage Learning, 2017.

EBBING, Darrell D.; WRIGHTON, Mark S. *General Chemistry*. 2nd ed. Boston: Houghton Mifflin Co., 1987.

EINSTEIN, Albert; METZ, André. A propos de «La Déduction Relativiste» de M. Émile Meyerson. *Revue Philosophique de la France et de l'Étranger*, **105**: 161-166, 1928.

FARRAR, Wilfred Vernon. Nineteenth-century speculations on the complexity of the chemical elements. *The British Journal for the History of Science*, **2** (4): 297-323, 1965.

FRUTEAU DE LACLOS, Frédéric. Émile Meyerson: a bibliography. *Iyyun: The Jerusalem Philosophical Quarterly*, **52**: 255-266, 2003.

FRUTEAU DE LACLOS, Frédéric. Le bergsonisme, point aveugle de la critique bachelardienne du continuisme d'Émile Meyerson. Pp. 109-122, in WORMS, Frédéric; WUNENBURGER, Jean-Jacques (eds.). *Bachelard et Bergson. Continuité et discontinuité*. Paris: Presses Universitaires de France, 2008.

GUERLAC, Henry. Chemistry as a branch of physics: Laplace's collaboration with Lavoisier. *Historical Studies in the Physical Sciences*, **7**: 193-276, 1976.

GUYTON DE MORVEAU, Louis Bernard; LAVOISIER, Antoine-Laurent de; BERTHOLLET, Claude-Louis; FOURCROY, Antoine François de. *Méthode de nomenclature chimique*. Proposée par MM. de Morveau, Lavoisier, Bertholet, et de Fourcroy; on y a joint un nouveau système de caractères chimiques, adaptés à cette nomenclature, par MM. Hassenfratz, Adet. Paris: Cuchet, 1787.

HAMERLA, Ralph Richard. Edward Williams Morley and the atomic weight of oxygen: the death of Prout's hypothesis revisited. *Annals of Science*, **60**: 351–372, 2003.

HAVEN, Kendall. *100 Greatest Science Discoveries of All Time*. Westport: Libraries Unlimited, 2007.

HILLMAN, Owen N. Émile Meyerson on scientific explanation. *Philosophy of Science*, **5** (1): 73-80, 1938.

HOLMES, Frederic L. Lavoisier the experimentalist. *Bulletin for the History of Chemistry*, **5**: 24-31, 1989.

HOLTON, Gerald. *Introduction to Concepts and Theories in Physical Science*. Cambridge MA: Addison-Wesley, 1952.

HOLTON, Gerald. What historians of science and science educators can do for one another. *Science & Education*, **12** (7): 603-616, 2003.

HOLTON, Gerald James; BRUSH, Stephen G. *Physics, the Human Adventure*. New Brunswick: Rutgers University Press, 2001.

HOOYKAAS, Reijer. *Fact, Faith and Fiction in the Development of Science*. The Gifford Lectures Given in the University of St Andrews 1976. Dordrecht: Springer Science, 1999.

IHDE, Aaron John. *The Development of Modern Chemistry*. New York: Dover, 1984.

KRAGH, Helge. The aether in late nineteenth century chemistry. *Ambix*, **36** (2): 49-65, 1989.

KUHN, Thomas S. *The Structure of Scientific Revolutions*. 50th Anniversary Edition. Chicago: University of Chicago Press, 2012.

LANDOLT, Hans Heinrich. Untersuchen über etwaige Änderungen des Gesamtgewichtes chemisch sich umsetzender Körper. *Sitzungsberichte der königl preussische Akademie der Wissenschaften zu Berlin*, **1**: 301-334, 1893.

LANDOLT, Hans Heinrich. Untersuchungen über die fraglichen Änderungen des Gesamtgewichtes chemisch sich umsetzender Körper. Zweite Mitteilung.

*Sitzungsberichte der königl preussische Akademie der Wissenschaften zu Berlin*, **14**: 266-298, 1906.

LANDOLT, Hans Heinrich. Untersuchungen über die fraglichen Änderungen des Gesamtgewichtes chemisch sich umsetzender Körper. Dritte Mitteilung. *Sitzungsberichte der königl Preussische Akademie der Wissenschaften zu Berlin*, **16**: 354-387, 1908.

LANDOLT, Hans Heinrich. Über die Erhaltung der Masse bei chemischen Umsetzungen. *Abhandlungen der königlich preussische Akademie der Wissenschaften zu Berlin. Physikalische-mathematische Classe*, Abhandlung 1: 1-158, 1910.

LAUE, Max von. Inertia and energy. Vol. 2, pp. 503-533, in SCHILPP, Paul Arthur (ed.). *Albert Einstein, philosopher-scientist*. New York: Harper and Brothers, 1959.

LAUGIER, Sandra. Science and realism: the legacy of Duhem and Meyerson in contemporary American philosophy of science. Pp. 91-140, in: BRENNER, Anastasios; GAYON, Jean (eds.). *French studies in the philosophy of science: contemporary research in France*. Dordrecht: Springer, 2009.

LAVOISIER, Antoine-Laurent de. *Résultats d'un ouvrage intitulé, De la richesse territoriale du royaume de France*. Paris: Imprimérie Nationale, 1791.

LAVOISIER, Antoine-Laurent de. *Traité élémentaire de chimie, présenté dans un ordre nouveau et d'après les découvertes modernes*. Paris: chez Cuchet, 1789. 2 vols.

LAVOISIER, Antoine-Laurent de. *Elements of Chemistry: In a New Systematic Order, Containing All the Modern Discoveries*. Transl. Robert Kerr. Third edition. Edinburgh: William Creech, 1796.

LAVOISIER, Antoine-Laurent de. *Oeuvres de Lavoisider. Tome II. Mémoires de chimie et de physique*. Paris: Imprimérie Impériale, 1862.

LECOURT, Dominique. *L'épistémologie historique de Gaston Bachelard*. Paris: Vrin, 2002.

LEICESTER, Henry M. Lomonosov's views on combustion and phlogiston. *Ambix*, **22** (1): 1-9, 1975.

MARION, Jerry. *Physical Science in the Modern World*. 2nd edition. Saint Louis: Elsevier Science, 2014.

MARTINS, Roberto de Andrade. Os experimentos de Landolt sobre a conservação da massa. *Química Nova*, **16** (5) : 481-490, 1993.

MARTINS, Roberto de Andrade. Eponyms as a stumbling block in the way of an adequate history of science. Paper 64, pp. 1-9, in: *IHPST 13th Biennial International Conference*. Rio de Janeiro, IHPST, 2015.

MARTINS, Roberto de Andrade. An educational blend of pseudohistory and history of science and its application in the study of the discovery of electromagnetism. Pp. 277-292, *in*: PRESTES, Maria Elice Brezinski; SILVA, Cibelle Celestino (eds.). *Teaching science with context: historical, philosophical, and sociological approaches*. Berlin: Springer, 2018.

MARTINS, Roberto de Andrade. Aspectos apriorísticos da ciência: Lavoisier e a conservação da massa nas reações químicas. Pp. 11-51, in: SILVA, Ana Paula Bispo da; MOURA, Breno Arsioli (orgs.). *Objetivos humanísticos, conteúdos científicos: contribuições da história e da filosofia da ciência para o ensino de ciências*. Campina Grande: EDUEPB, 2019 (a).

MARTINS, Roberto de Andrade. Émile Meyerson and mass conservation in chemical reactions: *a priori* expectations versus experimental tests. *Foundations of Chemistry*, **21** (1): 109-124, 2019 (b).

MARTINS, Roberto de Andrade; MARTINS, Lilian Al-Chueyr Pereira. Lavoisier e a conservação da massa. *Química Nova*, **16** (3): 245-256, 1993.

MEYERSON, Émile. Jean Rey et la loi de la conservation de la matière. *La Revue Scientifique de la France et de l'Étranger*, **33**: 299-303, 1884.

MEYERSON, Émile. Théodore Turquet de Mayerne et la découverte de l'hydrogène. *La Revue Scientifique de la France et de l'Étranger*, **42**: 665-670, 1888.

MEYERSON, Émile. La coupellation chez les anciens Juifs. *La Revue Scientifique de la France et de l'Étranger*, **47**: 756-758, 1891.

MEYERSON, Émile. *Identité et réalité*. Paris: Félix Alcan, 1908.

MEYERSON, Émile. *Identité et réalité*. 2nd edition. Paris: Félix Alcan, 1912.

MEYERSON, Émile. *De l'explication dans les sciences*. 2 vols. Paris: Payot, 1921.

MEYERSON, Émile. *La déduction relativiste*. Paris: Payot, 1925.

MEYERSON, Émile. *Identité et réalité*. 3rd edition. Paris: Félix Alcan, 1926.

MEYERSON, Émile. *Identidad y realidad*. Translated by Joaquím Xirau. Madrid: Editorial Reus, 1929.

MEYERSON, Émile. *Identity and Reality*. Translated by Kate Loewenberg. London: George Allen & Unwin, 1930 (a).

MEYERSON, Émile. *Identität und Wirklichkeit*. Translated by Leon Lichtenstein. Leipzig: Akademische Verlagsgesellschaft, 1930 (b).

MEYERSON, Émile. *Du cheminement de la pensée*. 3 vols. Paris: Alcan, 1931.

MEYERSON, Émile. Réel et déterminisme dans la physique quantique. Paris: Hermann, 1933.

MEYERSON, Émile. *Essais*. Paris: J. Vrin, 1936.

MEYERSON, Émile. *Explanation in the sciences*. Translated by Mary-Alice Sipfle and David A. Sipfle. Dordrecht: Springer, 1991.

MEYERSON, Émile. *Lettres françaises*. Edited by Bernadette Bensaude-Vincent and Eva Telkes-Klein. Paris: CNRS, 2009.

MEYERSON, Émile. *Mélanges. Petites pièces inédites*. Edited by Eva Telkes-Klein and Bernadette Bensaude-Vincent. Paris: Honoré Champion, 2011.

MORRIS, Robert J. Lavoisier and the caloric theory. *The British Journal for the History of Science*, **6** (1): 1-38, 1972.

PATY, Michel. Masse (de Newton à Einstein). Pp. 613-616, in: LECOURT, Dominique (ed.). *Dictionnaire d'Histoire et de Philosophie des Sciences*. Paris: Presses Universitaires de France, 1999.

PERRAUDIN, Jean François. A non-Bergsonian Bachelard. *Continental Philosophy Review* **41**(4): 463-479, 2008.

POMPER, Philip. Lomonosov and the discovery of the law of the conservation of matter in chemical transformations. *Ambix*, **10** (3): 119–127, 1962.

REY, Jean. *Essays of Jean Rey, doctor of medicine, on an enquiry into the cause wherefore tin and lead increase in weight on calcination* (1630). Edinburgh: The Alembic Club, 1904.

ROCKE, Alan. Letters to the Editor. *Newsletter of the History of Science Society*, **37** (2): 3, 2008.

ROSS, Sage. Wikipedia and the History of Science. *Newsletter of the History of Science Society*, **37** (1): 1, 6, 2008.

SCHELER, Lucien. Lavoisier et la régie des poudres. *Revue d'Histoire des Sciences*, **26** (3): 193-222, 1973.

SIEBRING, Barteld Richard; SCHAFF, Mary Ellen. *General Chemistry*. Belmont: Wadsworth, 1980.

SIEGFRIED, Robert. Lavoisier's table of simple substances: Its origin and interpretation. *Ambix*, **29** (1): 29-48, 1982.

SIEGFRIED, Robert. Lavoisier and the conservation of weight principle. *Bulletin for the History of Chemistry*, **5**: 18-24, 1989.

SIMONS, Massimiliano. The many encounters of Thomas Kuhn and French epistemology. *Studies in History and Philosophy of Science*, Part A, **61**: 41-50, 2017.

SMITH, John Russell. *Persistence and Periodicity: A Study of Mendeleev's Contribution to the Foundations of Chemistry*. PhD dissertation. London: University of London, 1976.

SOULIÉ, Stéphan Soulié. L'intégration d'Émile Meyerson à la communauté philosophique, le rôle de Xavier Léon et du réseau de la Revue de métaphysique et de morale. *Corpus, Revue de Philosophie*, **58**:129-142, 2010.

TELKES-KLEIN, Eva. Émile Meyerson, d'après sa correspondance. Une première ébauche. *Revue de Synthèse*, **125** (1): 197–215, 2004.

TELKES-KLEIN, Eva. Émile Meyerson, de la chimie à la philosophie des sciences. *Bulletin du Centre de Recherche Français à Jérusalem*, **18**: 107-116, 2007 (a).

TELKES-KLEIN, Eva. Meyerson dans les milieux intellectuels français dans les années 1920. *Archives de Philosophie*, **70** (3): 359-373, 2007 (b).

TELKES-KLEIN, Eva. La genèse d'*Identité et réalité* (1908) à travers une lettre d'Émile Meyerson à sa sœur. *Revue d'Histoire des Sciences*, **63** (1): 247-297, 2010.

TELKES-KLEIN, Eva. Émile Meyerson, philosophe des sciences et palestinophile (Lublin, 12 février 1859 – Paris, 2 décembre 1933). *Archives Juives*, **48** (2): 136-139, 2015.

USITALO, Steven. *The invention of Mikhail Lomonosov: a Russian national myth*. Boston: Academic Studies Press, 2013.

WETSHINGOLO, Ndjate-Lotanga. *La nature de la connaissance Scientifique: l'épistémologie Meyersonienne face à la critique de Gaston Bachelard*. Bern: Peter Lang, 1996.

WHITAKER, Robert D. An historical note on the conservation of mass. *Journal of Chemical Education*, **52** (10): 658-659, 1975.

# JEVONS AND THE ROLE OF ANALOGIES IN EMPIRICAL RESEARCH

Roberto de Andrade Martins

**Abstract**: Suppose a scientist discovers a new, unpredicted phenomenon (such as galvanism or ultraviolet radiation). How can one ascertain the causes, properties and laws of the phenomenon? How can one plan the investigation of the circumstances that affect the phenomenon, and of the effects that the new phenomenon could produce? If the phenomenon is completely unexpected and does not fit any previous theory, it is impossible to provide a theoretical prediction of its likely properties. In the empiricist tradition, therefore, the recommended method was to investigate all possibilities, because in such cases it is impossible to exclude *a priori* anything. William Stanley Jevons (1835-1882) provided a clear criticism of this method. It is impossible to investigate all possibilities, because they are boundless. Is it then impossible to plan the research of unexpected new phenomena? No. Jevons pointed out an alternative. According to Jevons, a scientist confronting a new, unexpected phenomenon, should compare it to other known phenomena to establish *analogies*. This comparison should allow the researcher to find out one or several known phenomena similar to the new one. This paper will present and discuss Jevons' proposal in the context of late 19th century methodology of science.
**Keywords**: scientific method; empirical research; analogy; Jevons, William Stanley

MARTINS, Roberto de Andrade. *Studies in History and Philosophy of Science II*. Extrema: Quamcumque Editum, 2021.

## 1. INTRODUCTION

In 1874 William Stanley Jevons (1835-1882) published the first edition of his book *The principles of science – a treatise on logic and scientific method*. In that work he discussed, among many other subjects, the strategies for the empirical inquiry of new phenomena in a pre-theoretical context. Suppose a scientist discovers a new, unpredicted phenomenon (such as galvanism or ultraviolet radiation). How can one ascertain the causes, properties and laws of the phenomenon? How can one plan the investigation of the circumstances that affect the phenomenon, and of the effects that the new phenomenon could produce?

If the phenomenon is completely unexpected and does not fit any previous theory, it is impossible to provide a theoretical prediction of its likely properties. In the empiricist tradition, therefore, the recommended method was to investigate all possibilities, because in such cases it is impossible to exclude *a priori* anything. This is the rule that can be found in Herschel's *A preliminary discourse on the study of natural philosophy*, for instance.

Jevons provided a clear criticism of this method. It is impossible to investigate all possibilities, because they are boundless. When a new unexpected phenomenon is discovered, any circumstance of the environment could, in principle, be essential for its production. It is impossible, however, to vary each of those factors. Besides that, even in the case of a small number of factors, a systematic empirical study would require the investigation of all possible combinations of the independent factors, and this would imply an overwhelming number of different tests.

A random choice of factors for investigation is also inadequate, of course, because one could miss the relevant factors and influences. Only in the case of a phenomenon predicted or suggested by some theory or hypothesis it is possible to select in advance the specific factors that are deemed relevant according to the theory or hypothesis under investigation.

Is it then impossible to plan the research of unexpected new phenomena? No. Jevons pointed out an alternative. According to him, a scientist confronting a new, unexpected phenomenon, should compare it to other known phenomena to establish *analogies*. This comparison should allow the researcher to find out one or several known phenomena similar to the new one. Then one should investigate whether the new phenomenon has properties equivalent to those of the known related phenomena.

Notice that the use of analogies to guide empirical research is not equivalent to the hypothetical-deductive method. An analogy does not imply a provisional belief. To use an analogy it is not necessary to assume that the phenomena are of the same nature: analogies between gravitation, electricity and magnetism have guided the experimental research of those phenomena without any assumption that they had equal causes. In the 18th century it was known that gravitational attraction was proportional to the mass and to the inverse square distance. By analogy, according to Jevons' methodological rule, one should investigate whether electricity and magnetism had equivalent properties.

It seems that Jevons' proposal was a new one. Jevons himself referred to Jeremy Bentham (*Essay on logic*) as his source of inspiration, but Bentham did not propose anything similar to Jevons' ideas. It also seems that his contemporaries did not understand Jevons' ideas. George Gore, for instance, read, commented and praised Jevons' book, but in his book *The art of scientific discovery* he claimed that the correct attitude of the scientist facing new phenomena should be to investigate all possibilities.

This paper will present and discuss Jevons' proposal in the context of late 19th century methodology.[1]

---

[1] This essay was written for presentation at the 11th International Congress of Logic, Methodology and Philosophy of Science [Section 16. History of Logic, Methodology, and Philosophy of Science],

## 2. THE PROBLEM: PLANNING EXPERIMENTS WITHOUT A THEORY

When scientific disciplines attains a high degree of development and there are working theories available, most of the experimental research is guided by those theories. However, in some special cases, a phenomenon is studied in a pre-theoretical environment. How can those inquiries be guided? How is it possible to plan observations and experiments when no theory is available?

It might seem that such a situation only occurs when science is in its infancy, but it can happen even within well-developed sciences. Sometimes an unexpected phenomenon is discovered. There are several well-known historical instances of chance discoveries in physics: the discovery of polarisation of light by refraction, the discovery of ultraviolet light, etc. It often occurs that the new phenomenon will be understood in the context of existing theories, even if it was not predicted. In that case, only the discovery of the new phenomenon occurs by chance – its further investigation is guided by theory. There are other cases, however, when the new phenomenon does not fit any existing theory.

Let us consider one famous instance: In 1895 Wilhelm Conrad Röntgen was studying electric discharges in vacuum tubes, when he noticed that a nearby fluorescent plate became bright. The unexpected phenomenon called his attention, and its study led to the discovery of a new kind of invisible penetrating radiation, with peculiar properties. From the very beginning of Röntgen's investigation, it became clear that the new radiation could not be explained by existing theories – it was a puzzle, and was the reason why it was called "X rays".

In this case, as in several other experimental discoveries in science, the researcher was not attempting to test any theory. In investigating something that had not been predicted – something

---

Cracow, Poland, 20-26 August 1999. It is published here for the first time. Part of its content had appeared in Portuguese (Martins, 1998).

that seemed completely new – it would impossible to use theories to plan his experiments. How should a scientist conduct his inquiry in those circumstances? Should he follow an inductive method? Should he do random experiments and observations? This was one of the problems William Stanley Jevons (1835-1882) addressed in his book *The principles of science – a treatise on logic and scientific method* (1874).

## 3.  JEVONS ON SCIENTIFIC METHOD

William Stanley Jevons[2] is well known for his books on Economy and Logic.[3] His work *The principles of science* is his best known contribution to epistemology and scientific method. About half of this book deals with logic: arguments, probability, deduction, induction, etc.[4] The second half of the treatise (starting from book IV – *Inductive investigation* – onward) deals

---

[2] There are few biographical studies on Jevons. See Gridgeman (1970) and other works referred there.

[3] Jevons' main works were: *A serious fall in the value of gold* (London, 1863); *Pure logic, or the logic of quality appart from quantity* (London, 1864; second edition in 1890); *The coal question* (London, 1865); *The substitution of similars* (London, 1869; second edition in 1890); *Elementary lessons in logic* (London, 1870); *The theory of political economy* (London, 1871; second edition in 1879); *The principles of science* (London, 1874; second edition in 1877); *Money and the mechanism of exchange* (London, 1875); *Primer on political economy* (London, 1878); *Studies and exercises in deductive logic* (London, 1880); *The State in relation to labour* (London, 1882); and the posthumous works *Methods of social* reform (London, 1883); *Investigations in currency and finance* (London, 1884); *Letters and journal of W. Stanley Jevons* (London, 1886, editado por sua esposa) and *The principles of economics* (London, 1905). Besides those books, Jevons published many articles on several subjects, including meteorology – a subject that strongly attracted him.

[4] This is the best known of Jevons' book. Most historians of philosophy and scientific method only discuss this part of his work. See, for instance, Madden (1966).

with the practice of scientific research. Most of the instances mentioned by Jevons are taken from physics, as was usual in the 19th century.

A large part of Jevons' book is dedicated to the analysis of experimental work. He was well aware that scientists do not follow and should not attempt to follow a Baconian "inductive method" because the bare accumulation of facts does not lead to the development of science. Experimental research is grounded upon theories, hypotheses and analogies, it begins with problems and questions, because without a previous conceptualisation it is impossible to plan an experiment.

This paper will deal with a specific point of Jevons' work: his ideas on the role of analogy in pre-theoretical scientific discovery. This specific point has not called the attention of historians, but it is an essential aspect of Jevons' thought. Indeed, Jevons regarded analogy or the comparison of similars as the fundamental principle of reasoning:

> In 1866 what he regarded as the great and universal principle of all reasoning dawned upon him; and in 1869 he published a sketch of this fundamental doctrine under the title of *The substitution of similars*. He expressed the principle in its simplest form as follows: "Whatever is true of a thing is true of its like," and he worked out in detail its various applications. (Hutchinson, 1947, p. 31)

> [...] The germ of his logical principles of the substitution of similars may be found in the view which he propounded in another letter written in 1861, that "philosophy would be found to consist solely in pointing out the likeness of things". (Hutchinson, 1947, p. 30)

> Jevons' logic of inference was dominated by what he called the substitution of similars, which expressed "the capacity of mutual replacement exiting in any two objects which are like or equivalent to a sufficient degree." This became for him "the great and universal principle of reasoning" from which "all logical processes seem to arrange

themselves in simple and luminous order". (Gridgeman, 1970, p. 105)

Although this was recognised as a central idea in Jevons' logical work, hitherto the role ascribed by Jevons to the substitution of similars in the scientific method has not been emphasised as it deserves.

Jevons' ideas on this subject are not presented in a closely-knit form in any chapter of his book. The account presented here is a *reconstruction* of Jevons' ideas, using statements scattered throughout his work.

## 4. JEVONS AND THE ROLE OF CHANCE IN SCIENTIFIC DISCOVERY

Jevons admitted that the discovery of a new scientific phenomenon might be due to accident or chance:

> No small part of the experience actually employed in science is acquired without any distinct purpose. We cannot use the eyes without gathering some facts which may prove useful. A great science has in many cases risen from an accidental observation. (Jevons, 1877, p. 399)[5]

Jevons presented several instances of accidental discovery: Bartholinus and Iceland spar, Galvani and the frog's leg, etc. Those are cases where the researcher was not looking for the phenomenon he discovered. Historical examples can show that chance discoveries do occur. However, Jevons not only presented historical cases: he attempted to justify the *necessity* of accidents for the empirical discovery of new phenomena:

> As a general rule we shall not know in what direction to look for a great body of phenomena widely different from those familiar to us. Chance then must give us the starting

---

[5] All references in this paper are to the second edition (1877) of *The principles of science.*

point; but one accidental observation well used may lead us to make thousands of observations in an intentional and organised manner, and thus a science may be gradually worked out from the smallest opening. (Jevons, 1877, p. 400)

I think that this is a correct argument. It would be absurd to think that a scientist could enter his laboratory and think: "Now, I am going to discover a new, unpredictable phenomenon". On the other hand, one cannot assume that the main part of scientific research is due to chance observations. The starting point might be a chance discovery, but after that, scientific investigation is not a random work.

Even chance discoveries are not completely random: it can only occur when the observer has the suitable knowledge allowing him to recognise that the observed fact is meaningful. Besides that, chance discoveries must be followed by an investigation of the new phenomenon, if they are to become useful, and that requires scientific training. According to Jevons:

> If we must attempt to draw a conclusion concerning the part which chance plays in scientific discovery, it must be allowed that it more or less affects the success of all inductive investigation, but becomes less important with the progress of science. Accident may bring a new and valuable combination to the notice of some person who had never expressly searched for a discovery of the kind, and the probabilities are certainly in favour of a discovery being occasionally made in this manner. But the greater the tact and industry with which a physicist applies himself to the study of nature, the greater is the probability that he will meet with fortunate accidents, and will turn them to good account. (Jevons, 1877, p. 532)

After the discovery of a new phenomenon it is necessary to think and talk about it. A scientist will not just inform "I have discovered a new *something*". He will ascribe a name or a simple description to the new phenomenon, and this early step will be guided by similarities to other known phenomena. Here, for the first time, Jevons introduced the use of *analogies*:

> When a phenomenon is of an unusual kind, we cannot even speak of it without using some analogy. Every word implies some resemblance between the thing to which it is applied, and some other thing, which fixes the meaning of the word. (Jevons, 1877, p. 522)

Historical examples are well known: "cells", cathodic "rays", electric "fluid", etc.

## 5. THE IMPOSSIBILITY OF EXAUSTIVE RANDOM INVESTIGATION

After the finding a new phenomenon, how should it be investigated?

In pre-theoretical research, the scientist must attempt to find the conditions related to the production, repetition or change of the phenomenon – that is, its empirical laws. This is the step Jevons called "inductive investigation":

> Our object in inductive investigations is to ascertain exactly the group of circumstances or conditions which being present, a certain other group of phenomena will follow. (Jevons, 1877, p. 416)

Can we study all circumstances or conditions that could affect a new phenomenon? Let us suppose that someone noticed for the first time that rubbing two sticks together turns them hot. An "exhaustive" account of the conditions of the observed phenomenon would have to include (Jevons, 1877, p. 416):
- the form, hardness, organic structure and all chemical qualities of the wood;
- the pressure and velocity of the rubbing
- the temperature, pressure, and all the chemical qualities of the surrounding air;

- the proximity of the earth with its attractive and electric powers;
- the temperature and other properties of the persons producing motion;
- the radiation from the sun, and to and from the sky; etc.

If we are facing a *new* phenomenon, how could we exclude the influence of any of those conditions, or even the influence of the colour of the clothes of the scientist? Only if the phenomenon is known it is possible to dismiss the influence of some of the circumstances.

> On *à priori* grounds it is unsafe to assume that any one of these circumstances is without effect, and it is only by experience that we can single out those precise conditions from which the observed heat of friction proceeds. (Jevons, 1877, pp. 416-7)

This is the first difficulty of experimental research in the pre-theoretical situation: there are infinite circumstances that could (in principle) affect the phenomenon and it is impossible to study infinite circumstances.

Even if it were possible to select a finite number of conditions that could affect the phenomenon, a second problem would arise: "The great difficulty of experiment arises from the fact that we must not assume the conditions to be independent" (Jevons, 1877, pp. 417). Suppose we want to test the influence of four factors *A*, *B*, *C*, *D* upon the phenomenon *P*. Suppose that **A** stands for the presence of factor *A*, and **a** stands for its absence. It would be necessary to observe whether *P* occurs or does not occur under all possible combinations such as **ABCD**, **aBCD**, **AbCD**, **abCD**, **ABcD**, etc.

> The effect of the absence of each condition should be tried both in the presence and absence of every other condition, and every selection of those conditions. Perfect and exhaustive experimentation would, in short, consist in examining natural phenomena in all their possible combinations and registering

all relations between conditions and results which are found capable of existence. (Jevons, 1877, pp. 417-8)

Is it *possible* to do this kind of exhaustive experimental analysis? Jevons remarked that even in the case of a finite number of conditions, the systematic research of all combinations is impossible, because the number of cases to be tested would increase according to an exponential law:

> The reader will perceive, however, that such exhaustive investigation is practically impossible, because the number of requisite experiments would be immensely great. Four antecedents only would require sixteen experiments; twelve antecedents would require 4096, and the number increases as the powers of two. [...] It is at this point that logical rules and forms begin to fail in giving aid. The logical rule is – Try all possible combinations; but this being impracticable, the experimentalist necessarily abandons strict logical method, and trusts to his own insight. (Jevons, 1877, p. 418)

Is this "insight" something that defies any understanding? According to Jevons there are no precise rules that could guide the experimenter, in this case:

> This work of inductive investigation cannot be guided by any system of precise and infallible rules, like those of deductive reasoning. There is, in fact, nothing to which we can apply rules of method, because the laws of nature must be in our possession before we can treat them. If there were any rule of inductive method, it would direct us to make an exhaustive arrangement of facts in all possible orders. (Jevons, 1877, p. 504)

> We may be obliged to trust to the casual detection of coincidences in those branches of knowledge where we are deprived of the aid of any guiding notions; but a little reflection will show the utter insufficiency of haphazard experiment, when applied to investigations of a complicated

nature. [...] When considering the subject of combinations and permutations, it became apparent that we could never cope with the possible variety of nature. An exhaustive examination of the possible metallic alloys, or chemical compounds, was found to be out of the question (Jevons, 1877, p. 505)

Jevons' argument might appear so straightforward that it should have occurred to everyone else. That was not the case. Let us first consider the opinion of another famous 19th century methodologist: John Herschel. According to Herschel, when a new fact is described it would be necessary to include *all the circumstances* of its occurrence, and then one would need to study which are relevant or otherwise:

> The circumstances, then, which accompany any observed fact, are main features in its observation, at least until it is ascertained by sufficient experience what circumstances have nothing to do with it, and might therefore have been left unobserved without sacrificing *the fact*. In observing and recording a fact, therefore, altogether new, we ought not to omit any circumstance capable of being noted, lest some one of the omitted circumstances should be essentially connected with the fact, and its omission should, therefore, reduce the implied statement of a *law of nature* to the mere record of an *historical event*. (Herschel, 1966, p. 120, § 111)

So, Jevons was not describing "common sense epistemology". Although his views may accord with our "common sense", they were not obvious or consensual. Nevertheless, Jevons did not present his approach as a new analysis. He stated that his opinion was the result of a gradual reaction against the method prescribed by Francis Bacon:

> It would be an interesting work, but one which I cannot undertake, to trace out the gradual reaction which has taken place in recent times against the purely empirical or Baconian theory of induction. Francis Bacon, seeing the futility of the

scholastic logic, which had long been predominant, asserted that the accumulation of facts and the orderly abstraction of axioms, or general laws from them, constituted the true method of induction. [...]

Nevertheless Bacon's method, as far as we can gather the meaning of the main portions of his writings, would correspond to the process of empirically collecting facts and exhaustively classify them, to which I alluded[6]. The value of this method may be estimated historically by the fact that it has not been followed by any of the great masters of science (Jevons, 1877, pp. 506-7)

In principle, if we are facing a new phenomenon, no possibility can be excluded *a priori*, but scientists do manage to investigate new phenomena and they never analyse all possibilities. How do they do that?

One might immediately recall the hypothetical-deductive method: instead of investigating *all* possibilities, the scientist formulates hypotheses and only tests the consequences of those hypotheses. In the Introduction he wrote for the Dover edition of *The principles of science* Nagel called the attention of the readers to Jevons' approach to the use of hypotheses in scientific research (Nagel, 1958, pp. xlix-li). However, Nagel did not pay attention to another question: How are hypotheses formulated? In the pre-theoretical context there is an infinite number of possible hypotheses – indeed, it is possible to frame hypothetical relations that would contemplate all possible combinations of circumstances around the phenomenon. If the choice of hypotheses is not blind or random, what can guide their choice?

---

[6] Jevons described here a common view on Bacon's method. It is possible however to find in Bacon's work another less conspicuous approach, suggesting the careful use of analogies: "There is no proceeding in invention of knowledge but by similitude" (see Park, 1984, p. 297).

## 6. ANALOGY AS A SOURCE OF WORKING HYPOTHESES

Jevons clearly pointed out that the pre-theoretical investigation should be guided by analogies, grounded upon the scientist's experience and "intuition":

> The reader will perceive, however, that such exhaustive investigation is practically impossible [...]. The result is that the experimenter has to fall back upon his own tact and experience in selecting those experiments which are most likely to yield him significant facts. It is at this point that logical rules and forms begin to fail in giving aid. The logical rule is – Try all possible combinations; but this being impracticable, the experimentalist necessarily abandons strict logical method, and trusts to his own insight. Analogy, as we shall see, gives some assistance, and attention should be concentrated on those kinds of conditions which have been found important in like cases. But we are now entirely in the region of probability, and the experimenter, while he is confidently pursuing what he thinks the right clue, may be overlooking the one condition of importance. (Jevons, 1877, p. 418)

> As natural science progresses, physicists gain a kind of insight and tact in judging what qualities of a substance are likely to be concerned in any class of phenomena. (Jevons, 1877, p. 422)

There is no safe rule to exclude any given condition or combination of conditions. However, it is impossible to study all the infinite possibilities. The researcher must choose, and he does choose, taking into account his "tact", "experience", "insight", using analogies to guide his work. Instead of contemplating facts with his mind void of ideas, as required by the Baconian method, it is necessary to use working hypotheses:

In later years Professor Huxley has strongly insisted upon the value of hypothesis. When he advocates the use of "working hypotheses" he means no doubt that any hypothesis is better than none, and that we cannot avoid being guided in our observations by some hypothesis or other. (Jevons, 1877, p. 509)

Of course, the use of hypotheses is not a new idea, but Jevons' analysis of the origin of hypotheses seems new.

## 7.  JEVONS' CONCEPT OF 'ANALOGY'

What does it mean to say that two phenomena present an analogy? Jevons did not present an explicit definition of this concept in *The principles of science*, but from the very beginning of the book he emphasises the role of analogies and comparisons in all kinds of inferences:

The fundamental action of our reasoning faculties consists in inferring or carrying to a new instance of a phenomenon whatever we have previously known of its like, analogue, equivalent or equal. Sameness or identity presents itself in all degrees, and is known under various names; but the great rule of inference embraces all degrees, and affirms that *so far as there exists sameness, identity or likeness, what is true of one thing will be true of the other*. (Jevons, 1877, p. 9)[7]

Analogy is a special (imperfect) case of the *substitution of similars*, which Jevons regarded as the most important rule of inference:

The one supreme rule of inference consists, as I have said, in the direction to affirm of anything whatever is known of its like, equal or equivalent. The *Substitution of Similars* is a phrase which seems aptly to express the capacity of mutual replacement existing in any two objects which are like or

---

[7] "The universal principle of all reasoning, as I have asserted, is that which allows us to substitute like for like" (Jevons, 1877, p. 162).

equivalent to a sufficient degree. It is matter for further investigation to ascertain when and for what purposes a degree of similarity less than complete identity is sufficient to warrant substitution. (Jevons, 1877, p. 17)

At another place Jevons elucidated what it means to reason by analogy:

> In reasoning by analogy, the, we observe that two objects A B C D E ... and A' B' C' D' E' ... have many like qualities, as indicated by the identity of the letters, and we infer that, since the first has another quality, X, we shall discover this quality in the second case by sufficiently close examination. As Laplace says, – "Analogy is founded on the probability that similar things have causes of the same kind, and produce the same effects. The more perfect this similarity, the greater is this probability". (Jevons, 1877, p. 597)

In the case of a chance discovery, the unexpected observation of a new phenomenon will evoke, by association, several similar phenomena. Analogy will guide the search for other features of the new phenomenon.

## 8.  THE ANALYSIS OF SIMILARITIES

According to Jevons, when the experimenter is confronted with a new phenomenon, he should use analogic reasoning to suggest hypotheses; then, he should test them:

> It is before the glance of the philosophic mind that facts must display their meaning, and fall into logic order. The natural philosopher must therefore have, in the first place, a mind of impressionable character, which is affected by the slightest exceptional phenomenon. His associating and identifying powers must be great, that is, a strange fact must suggest to his mind whatever of like nature has previously come within his experience. His imagination must be active, and bring before his mind multitudes of relations in which the unexplained facts may possibly stand with regard to each

other, or to more common facts. Sure and vigorous powers of deductive reasoning must them come into play, and enable him to infer what will happen under each supposed condition. Lastly, and above all, there must be the love of certainty leading him diligently and with perfect candour, to compare his speculations with the test of fact and experiment. (Jevons, 1877, p. 577)

Hypotheses used in pre-theoretical investigation cannot be deduced from any previous set of established propositions. They are *suggested* by analogy, from former knowledge. After that, it is necessary to test the hypotheses.

If the views upheld in this work be correct, all inductive investigation consists in the marriage of hypothesis and experiment. When facts are in our possession, we frame an hypothesis to explain their relations, and by the success of this explanation is the value of the hypothesis to be judged. In the invention and treatment of such hypotheses, we must avail ourselves of the whole body of science already accumulated, and when once we have obtained a probable hypothesis, we must not rest until we have verified it by comparison with new facts. We must endeavour by deductive reasoning to anticipate such phenomena, especially those of a singular and exceptional nature, as would happen if the hypothesis be true. Out of the infinite number of experiments which are possible, theory[8] must lead us to select those critical ones which are suitable for confirming or negating our anticipation. (Jevons, 1877, p. 504)

The true course of inductive procedure is that which has yielded all the more lofty results of science. It consists in *Anticipating Nature*, in the sense of forming hypotheses as to the laws which are probably in operation; and then observing whether the combinations of phenomena are such as would follow from the laws supposed. The investigator begins with

---

[8] In a pre-theoretical context, the "probable hypotheses" are obtained by analogy, according to Jevons.

facts and ends with them. He uses facts to suggest probable hypotheses; deducing other facts which would happen if a particular hypothesis is true, he proceeds to test the truth of his notion by fresh observations. If any result prove different from what he expects, it leads him to modify or to abandon his hypothesis; but every new fact may give some new suggestion as to the laws in action. Even if the result in any case agrees with his anticipations, he does not regard it as finally confirmatory of his theory, but proceeds to test the truth of the theory by new deductions and new trials. (Jevons, 1877, p. 509)

In such a process the investigator is assisted by the whole body of science previously accumulated. He may employ analogy, as I shall point out, to guide him in the choice of hypotheses. The manifold connections between one science and another give him clues to the kind of laws to be expected, and out of the infinite number of possible hypotheses he selects those which are, as far as can be foreseen at the moment, most probable. Each experiment, therefore, which he performs is that most likely to throw light upon his subject, and even if it frustrate his first views, it tends to put him in possession of the correct clue. (Jevons, 1877, pp. 509-510)

It is possible to read in those quotes just an apology of the hypothetical-deductive method. I would like to stress, however, Jevons' views on the *use of analogy in the invention of hypotheses*. Jevons himself acknowledged that previous authors had proposed this idea:

[...] As Boscovich truly said, we are to understand by hypotheses "not fictions altogether arbitrary, but suppositions conformable to experience or analogy". It follows that every hypothesis worthy of consideration must suggest some likeness, analogy, or common law, acting in two or more things. (Jevons, 1877, p. 512)

However, it seems that no other author, before or after Jevons, gave so much emphasis to analogy.

## 9.  UNCERTAINTY OF WORKING HYPOTHESES

Jevons was aware that hypotheses suggested by analogy are just working instruments and that they are open to error. However, they are useful and even necessary in the experimental investigation of nature:

> There can be no doubt that discovery is most frequently accomplished by following up hints received from analogy, as Jeremy Bentham remarked[9]. Whenever a phenomenon is perceived, the first impulse of the mind is to connect it with the most nearly similar phenomenon. If we could ever meet a thing wholly *sui generis*, presenting no analogy to anything else, we should be incapable of investigating its nature, except by purely haphazard trial. The probability of success by such a process is so slight, that it is preferable to follow up the faintest clue. As I have pointed out already (p. 418), the possible experiments are almost infinite in number, and very numerous also are the hypotheses upon which we may proceed. Now it is self-evident that, however slightly superior the probability of success by one course of procedure may be over another, the most probable one should always be adopted first. (Jevons, 1877, p. 629)

As hypotheses are uncertain, instead of fixing his mind upon a single one, the researcher should investigate a large number of hypotheses:

> It would be an error to suppose that the great discoverer seizes at once upon the truth, or has any unerring method of divining it. In all probability the errors of the great mind exceed in number those of the less vigorous one. Fertility of imagination and abundance of guesses at truth are among the first requisites of discovery; but the erroneous guesses must be many times as numerous as those which prove well founded. The weakest analogies, the most whimsical notions,

---

[9] Here Jevons referred to Bentham's *Essay on logic*: (Bentham, 1843, vol. 8, p. 276).

the most apparently absurd theories, may pass through the teeming brain, and no record remain of more than the hundredth part. There is nothing really absurd except that which proves contrary to logic and experience. The truest theories involve suppositions which are inconceivable, and no limit can really be placed to the freedom of hypothesis. (Jevons, 1877, p. 577)

Since the procedure of invention of hypotheses using analogies does not provide any warranty as to their certainty, the researcher should develop rigorous test of those hypotheses:

Summing up, then, it would seem as if the mind of the great discoverer must combine contradictory attributes. He must be fertile in theories and hypotheses, and yet full of facts and precise results of experience. He must entertain the feeblest analogies and the merest guesses at truth, and yet he must hold them as worthless till they are verified in experiment. When there are any grounds of probability he must hold tenaciously to an old opinion, and yet he must be prepared at any moment to relinquish it when a clearly contradictory fact is encountered. (Jevons, 1877, pp. 592-593)

In the specific case of pre-theoretical experimental investigation, the scientist should therefore frame several hypotheses concerning the properties of the new phenomenon, using analogies about known phenomena. Then, he should submit the hypotheses to a careful experimental investigation. In this way, before a *theory* is reached, the new phenomenon is progressively understood, because, according to Jevons, the general sense of explanation is the search for similarities between the new and the known (Jevons, 1877, p. 533).

Reasoning by analogy can suggest an indefinite number of properties that could be investigated. The associations that will actually occur to a particular researcher during this phase of his inquiry will depend on his background and on his current interests. Heath stated that, according to Jevons, it would be necessary to test all conceivable hypotheses (Heath, 1967, p.

261). That was not Jevons' ideal. According to him, a scientist only needs to check the particular conjectures that occurred to him from the analogies that spontaneously arose in his mind.

## 10. JEREMY BENTHAM ON ANALOGY AND DISCOVERY

It seems that Jevons' proposal concerning the use of analogy in scientific research was a new one. Jevons himself referred to Jeremy Bentham (*Essay on logic*) as his source of inspiration, but Bentham did not propose anything similar to Jevons' ideas. Jevons referred to the second section of chapter 10 (Bentham, 1843, pp. 275-279 – "On the art of invention") of Bentham's book. In this section Bentham describes "helps applicable to arts in general without exception or distinction", divided in 10 hints or aid to memory (*memento*). Two of those rules mention analogy:

> *Memento 5.* For means and instruments, employ analogy. *Analogias undique indagato.*
> *Memento 6.* In your look-out for analogies, for surveying that quarter of the field of thought and action to which the art in question belongs, employ the logical ladders made of nest of aggregates, placed in logical sub-alternation. *In analogiarum indagationes scalis logicis utere.* (Bentham, 1843, p. 276)

Bentham explained the use of analogy as an analysis of genera and species: whatever is true of a genus, should be true of each species belonging to that genus; and whatever is true of a species belonging to a genus *might be true* of the genus and of other species of the same genus.

> *Fifth and Sixth Mementos*: The mode and use of applying these *subalternation scales* are as follows, viz.:
> I. Application in the *descending line*.
> With the exception of such words as are names of individual objects, take any one of the material words that

present themselves as belonging to the subject, not being the name of an individual alone, this word will be the name of a *sort* of objects, the name, (say) of an aggregate. If the aggregate be the denomination of a *genus*, think of the several species which, by their respective names, present themselves as being contained under it. Whatsoever is predicated of the genus, will, in so far as it is truly predicated, be, with equal truth, predicable of all these several species.

II. Application in the *ascending line.*

In like manner look out for the name of the next superior genus; with reference to which, the genus in question is but a species, and observe, try, or conjecture, whether that which beyond doubt, has been found predicable with truth of the whole of this species, be, or promises to be, with like truth predicable of the whole, or any other part of the aggregate, designated by the name of that genus. (Bentham, 1843, p. 278)

The application of Bentham's "subalternation scale in the ascending line" might be regarded as a kind of analogical reasoning. However, immediately after the above quoted elucidation, Bentham added:

It is in the instance of the physical department of the field of thought and action, and more particularly to the chemical district of that department, that the applicability of this memento is most conspicuous. Upon every subject, try, or at least, think of trying, every operation; to every subject in the character of a menstruum, apply every subject in the character of a solvent, and so on. (Bentham, 1843, p. 278)

Now, Bentham proposed to test every possibility – an impossible method, according to Jevons. Not only did Bentham think that the use of analogical thinking would entail this endless search, but he even raised this combinatorial method to the status of a new rule:

*Memento 9. Quodlibet cum quolibet.* To everything forget not to apply anything. Suppose that of an indefinite multitude of objects, which in consideration of certain properties or qualities, in respect of which they are found or supposed to agree, and certain others, in respect of which they have been found or supposed to disagree, having all of them been placed in one or other of two classes, some article belonging to the one class has, with success, (i.e. with some new effect, which either has been found to be, or affords a prospect of being found to be, advantageous,) been applied, no matter in what manner, nor to what purpose in particular, to some article belonging to the other class; in like manner, frame a general resolution not to be departed from in any instance, but for some special cause, (applying to that instance,) to apply to each article belonging to the one class every article belonging to the other. (Bentham, 1843, p. 276)

Therefore, in Bentham's account an essential feature of Jevons' approach is lacking: the clear statement that it is impossible to investigate every possibility.

## 11. GORE AND THE METHOD OF EXPERIMENTAL INVESTIGATION

It seems that his contemporaries did not understand Jevons' ideas. Let us analyse one relevant case: Gore's reaction to Jevons' proposals.

In 1878 the chemist George Gore (1826-1908)[10] published his book *The art of scientific discovery or the general conditions and methods of research in physics and chemistry*. In the Preface to this work he cited twice, among other useful books, Jevons *Principles of science* (Gore, 1878, p. vi and p. x – footnote 2). He also acknowledged Jevons' help in correcting part of his book (*ibid.*, p. xiii). Besides that, in several parts of his work Gore approvingly cited Jevons' book, and never

---

[10] There is a short biographical notice on George Gore in the *Dictionary of Scientific Biography*: Jones, 1970.

criticised him. It seems, therefore, that Gore accepted Jevons' ideas.

At several points of his book Gore does present an interpretation of the scientific method that seems inspired by Jevons' book. He stressed the utility of comparison and analogy as an aid to discovery (Gore, 1878, pp. 327-331) and as one of the sources of hypotheses (*ibid.*, pp. 366-7). However, a few pages later, he claimed that the correct attitude of the scientist facing new phenomena should be to investigate all possibilities:

> The power, activity, and variety of the imagination may be considerably increased by practising the formation of hypotheses, in the manner already described, on every available opportunity. This practice may be greatly assisted by the use of a table of classified series of leading ideas of the various sciences, and associating each of these ideas in succession with that of the phenomenon under consideration, and then forming questions respecting it by asking in succession what effect will each have upon the particular phenomenon. The following fragment of such a table will show what I mean: – What will be the effect of gravity, pressure, motion, heat, light, electricity, magnetism, chemical affinity; and of varying time of action, direction, and strength of each of these; also the effect of conduction, radiation, refraction, reflection, and polarisation of heat upon it. And so on through all the chief phenomena of all the forces of nature in succession; and also asking what will be the effect of different classes of elementary substances, metals, metalloids, &c., and all the separate elementary substances and their compounds in succession. Instead of such a table, a copious index of any good book on physical and chemical science may be employed for the purpose. In this way even a student of science may suggest a large number of new questions respecting any phenomenon. (Gore, 1878, pp. 369-70)

The method described by Gore is the kind of blind combinatorial investigation that was criticised by Jevons. Thus, one of the essential features of Jevons' methodological analysis was overlooked by Gore.

## 12. THE EPISTEMOLOGICAL STATUS OF ANALOGY IN DISCOVERY

Up to this point I have attempted to describe Jevons' ideas and to stress his originality. Of course, I do think that Jevons' proposal is a relevant contribution to the scientific method – otherwise I would not have chosen to write on this subject. However, before closing this paper let me add some critical comments on Jevons' views on analogy.

From a historical point of view, the concept of 'analogy' was born in mathematics where it meant an equality between ratios or proportions (Lloyd, 1973, p. 60).[11] Afterward this word was used in several different senses (Hesse, 1967). Although there is a wide range of analogy concepts, let us assume the following statement as a reasonable account of most recent uses of this word:

- *Two objects A and B of any kind are analogous if there are parts, properties or relations that are similar or equal in both A and B (that is, if they have some equivalent features) and if, beside that, they have some difference.*

If two objects are analogous, this similarity *suggests* that they might have other equivalent features:

> The examination of likeness is useful with a view both to inductive arguments and to hypothetical reasonings, and also with a view to the rendering of definitions. [...] It is useful for hypothetical reasonings because it is a general opinion that among similars what is true of one is true also of the rest. If,

---

[11] Notice that even in ancient Greek thought analogy was also regarded as a method of suggesting explanations of natural phenomena (Lloyd, 1973, p. 63).

then, with regard to any of them we are well supplied with matter for discussion, we shall secure a preliminary admission that however it is in these cases, so it is also in the case before us [...] (Aristotle, *Topics*, book I, chapter 18, 108$^b$ 6-16)

'Reasoning by analogy' consists on inferring an unknown similarity between two objects, from a known analogy between them. Of course, reasoning by analogy is not demonstrative. What does it produce, then? Jevons supposed that it leads to *hypotheses*, and so does Carnap:

> The evidence known to us is the fact that individuals $b$ and $c$ agree in certain properties and, in addition, that $b$ has a further property; thereupon we consider the hypothesis that $c$ too has this property. (Carnap, 1962, p. 569)

Carnap and most of other authors regard the result produced by analogy as *likely or probable hypotheses*. The greater the initially known similarity between the two objects, the greater will be the probability of the hypothesis. As shown above, Jevons also suggested a probabilistic interpretation for reasoning by analogy.

Other authors – such as Norwood Hanson – who also accepted that analogy has an important role in the formulation of new hypotheses supposed that the process of discovery consisted on the formulation of *plausible* hypotheses: previous knowledge would lead a scientist to think that the hypothesis is probable (Hanson, 1958, p. 1074).

However, as shown by Mary Hesse, this concept is highly problematical. Any analogy involves not only similarities but also differences. Knowledge of those differences should also be taken into account when the probability of the hypothesis is to be evaluated, but it is very difficult to find out a viable way of doing this (Hesse, 1964).

Indeed, let us start from the supposition that $A$ and $B$ agree in the properties $m, n, p, q$ and disagree in the properties $t, v, w$.[12] Given this knowledge, what is the probability that $A$ and $B$ agree in every property? Of course, the probability is null, and therefore it is impossible to assume that $A$ and $B$ agree in a new property $x$.[13] Now, if the new property $x$ is completely independent of the properties $m, n, p, q, t, v, w$, it is impossible to find out the probability of $x$ belonging to $B$, given that it belongs to $A$. It seems impossible to frame an acceptable concept of probability that could be applied to analogous reasoning.

The difficulty of applying the concept of probability to the conclusions of arguments by analogy could be regarded as a fatal blemish of Jevons' view. However, most of his ideas might be retained if analogies are regarded under another point of view.

Instead of interpreting analogy as a kind of inference producing a *proposition* (something that might be regarded as true or false, or more or less probable), let us embrace another approach: analogy will be regarded as the source of *rules of action* of a special kind. I propose that reasoning by analogy in the context of pre-theoretical scientific research can be described by two rules:

- *Given a new phenomenon, it is desirable to establish analogies between the new phenomenon and other known phenomena.*

---

[12] In any real-world situation, two different objects will disagree in an infinite number of ways, but let us suppose that we only know that they disagree in a finite number of features.

[13] If we only knew that $A$ and $B$ agree in some properties, it would be possible, according to Carnap, to ascribe a probability between 0 and 1 to the statement that $A$ and $B$ agree in every property, and therefore the hypothesis that $A$ and $B$ would agree in a new property $x$ would be different from zero.

- *If we know that two phenomena A and B are analogous and we also know that A has a feature p and we do not know whether B has a similar feature or not, it is desirable to find out whether B has that feature or not.*

So, according to this proposal, analogical reasoning will not lead to state that *B* has the property *p*. It will lead to an *action*: the scientist should attempt to find out whether *B* has that property of otherwise. The result of reasoning by analogy would not produce *propositions*, in the logical sense, but *desiderata* that could guide research[14]. The analysis of similarities and differences between *A* and *B* shows that some information is lacking (we do not know whether *p* applies to *B* or not) and the researcher is led to fill this gap[15]. This is not equivalent to a hypothesis proper, because a hypothesis is a proposition that is regarded as probable. When a scientist is guided by the above-described rule of action, it is completely indifferent whether *p* applies to *B* or not. Analogy cannot lead a scientist to believe that "*p* applies to *B*" is true or probable and therefore he should not defend that he will likely find the property *p* to apply to *B*. However, without applying any truth-value or probability to the proposition "*p* applies to *B*", the above described rule directs the effort of the scientist to choose between "*p* applies to *B*" or "*p* does not apply to *B*".

The *logical* value of this rule of action is null, because the result of the reasoning by analogy would amount to a tautology: "either *p* applies to *B* or *p* does not apply to *B*". However, the *methodological* value of reasoning by analogy is very high,

---

[14] Reasoning by analogy could also be described as a way to arrive to *questions* such as this: "Does B have the property p?".

[15] A *desideratum* is the description of something that is useful or desirable, from the scientific point of view. It describes sufficient (but not necessary) conditions for ascribing positive scientific value to a scientific result. A general analysis of *desiderata* can be found in my PhD thesis: Martins, *Sobre o papel dos* desiderata *na ciência* (1987). See also: Martins, 1980; Martins, 1984.

because it will focus the attention of the scientist upon a few specific features of the phenomenon, instead of asking him to analyse an infinity of possibilities.

Notice that a *desideratum* is not a heuristic method. Heuristic procedures can help to find new analogies, but the rules presented here do not provide a way to find an analogy – they only state that it is *desirable* to establish analogies, that is, analogies have a positive scientific value.

Except for this feature – the epistemological status of analogical reasoning – I think that the view presented by Jevons is both a nice account of actual scientific practice[16] and a useful methodological guide for the experimental investigation of new phenomena in a pre-theoretical context.

## 13. FINAL REMARKS

This paper discussed the role of analogy in the process of scientific discovery and experimental investigation in a pre-theoretical context, following Jevons' work. The paper did not attempt to discuss all kinds of analogy, nor every kind of process of discovery and research. Its aim was not to discuss the whole of Jevons' contribution to scientific method and epistemology, but to deal with a specific point of his work.

Jevons' analysis suggest that the discovery of a completely new phenomenon is due to chance, but soon afterwards some assumptions must provide a guidance to the research – otherwise, observations and experiments would occur at random, and that would seldom lead to significant results. It is impossible to study all the factors that could possibly affect a give phenomenon, and hence the attention of the researcher should be focused upon a small number of features. This choice, in a pre-theoretical context, must be guided by analogies.

---

[16] In this paper I have not attempted to compare Jevons' analysis to the actual scientific pratice. In another paper, however, I have shown that the discovery and early investigation of X rays followed a method corresponding to Jevons' ideas (Martins, 1998).

Jevons' seminal work was not derived from former methodological accounts. It was at variance with Herschel's ideas, and was not derived from Bentham's work – notwithstanding Jevons' own remarks on his debt to that author. Jevons ideas were not straightforward and were not understood by contemporary authors, such as Gore.

Contrary to traditional analyses, this paper ascribes a peculiar epistemological status to analogies used in that kind of investigation: they are interpreted as originating *desiderata*, instead of probable hypotheses. According to this interpretation, analogies are not the source of probable or plausible hypotheses, but are used to focus the attention of the scientist upon a few features of the new phenomenon, in such a way as to avoid random or endless experimental inquiries.

## ACKNOWLEDGEMENT

The author is grateful to the Brazilian National Council for Scientific and Technological Development (CNPq) for supporting this research.

## BIBLIOGRAPHIC REFERENCES

BENTHAM, Jeremy. *Essay on logic*. Vol. 8, pp. 213-293 in BENTHAM, Jeremy. *Works*. Ed. John Bowning. 12 vols. Edinburgh: William Tait, 1843.

CARNAP, Rudolf. *The logical foundations of probability*. 2. ed. Chicago: The University of Chicago Press, 1962.

GORE, George. *The art of scientific discovery or the general conditions and methods of research in physics and chemistry*. London: Longmans, Green, and Co., 1878.

GRIDGEMAN, Norman T. Jevons, William Stanley. Vol. 7, pp. 103-107 in GILLISPIE, Charlton Couslton (ed). *Dictionary of scientific biography*. New York: Charles Scribners Sons, 1970.

HANSON, Norwood Russell. The logic of discovery. *The Journal of Philosophy* **55**: 1073-1089, 1958.

HEATH, P. L. Jevons, William Stanley. Vol. 4, pp. 260-261, in EDWARDS, Paul (ed.). *The encyclopaedia of philosophy*. New York: Macmillan & The Free Press, 1967.

HERSCHEL, John Frederick William. *A preliminary discourse on the study of natural philosophy*. New York: Johnson Reprint, 1966. (Sources of Science, n° 17)

HESSE, Mary. Analogy and confirmation theory. *Philosophy of Science* **31** (4): 319-327, 1964.

HESSE, Mary. Models and analogy in science. Vol. 5, pp. 354-359, *in*: EDWARDS, Paul (ed.). *The encyclopaedia of philosophy*. New York: Macmillan & The Free Press, 1967.

HUTCHINSON, Terence Wilmot. Jevons, William Stanley (1835-1882). Vol. 13, pp. 30-31 in *Encyclopaedia Britannica*. Chicago: Encyclopaedia Britannica, 1947.

JEVONS, William Stanley. *The principles of science – a treatise on logic and scientific method* [1877]. London: MacMillan, 1924.

JONES, Daniel P. Gore, George. Vol. 5, p. 474 in GILLISPIE, Charlton Couslton (ed). *Dictionary of scientific biography*. New York: Charles Scribners Sons, 1970.

LLOYD, G. E. R. Analogy in early Greek thought. Vol. 1, pp. 60-63, *in*: WIENER, Philip P. (ed.). *Dictionary of the history of ideas: studies of selected pivotal ideas*. New York: Charles Scribner's Sons, 1973.

MADDEN, Edward H. W. S. Jevons on induction and probability. Pp. 233-247, *in*: MADDEN, Edward H. (ed.). *Theories of scientific method: the Renaissance through the nineteenth century*. Seattle: The University of Washington Press, 1966

MARTINS, Roberto de Andrade. Abordagem axiológica da epistemologia científica. *Textos SEAF* **1** (2): 38-57, 1980.

MARTINS, Roberto de Andrade. A situação epistemológica da epistemologia. *Revista de Ciências Humanas, UFSC* **3** (5): 85-110, 1984.

MARTINS, Roberto de Andrade. *Sobre o papel dos* desiderata *na ciência*. Campinas: UNICAMP, 1987.

MARTINS, Roberto de Andrade. Jevons e o papel da analogia na arte da descoberta experimental: o caso da descoberta dos raios X e sua investigação pré-teórica. *Episteme* **3** (6): 222-249, 1998.

NAGEL, Ernest. Introduction to the Dover edition. Pp. xlv-liii in JEVONS, William Stanley. *The principles of science – a treatise on logic and scientific method.* New York: Dover, 1958.

PARK, Katharine. Bacon's "enchanted glass". *Isis* **75**: 290-302, 1984.

# WAVE MECHANICS, FROM LOUIS DE BROGLIE TO SCHRÖDINGER: A COMPARISON

Roberto de Andrade Martins

**Abstract**: Erwin Schrödinger's work on wave mechanics started in late 1925, stimulated by his study of Louis de Broglie's thesis. It is well known that in his initial attempts to formulate a quantum theory of the atom Schrödinger tried to develop a relativistic theory, following de Broglie's ideas, and only afterwards he looked for a non-relativistic wave equation. It is straightforward to derive the wave equation corresponding to de Broglie's phase waves. both in the relativistic and non-relativistic realms. In the case of his relativistic attempt, Schrödinger did indeed follow a simple approach, using de Broglie's theory. In the non-relativistic approach, he attempted to produce an independent derivation of the wave equation, following several different lines, instead of using de Broglie's results in the classical limit. This paper analyses Schrödinger's derivations of the wave equation, showing the differences and similarities between his theory and de Broglie's. It will be shown that, although it is formally possible to derive the wave equation from de Broglie's theory, there is an incompatibility between the two theories: it would be impossible to make any sense of de Broglie's ideas in the case of the rigid rotator, for instance. Schrödinger's approach was, in this sense, independent and incompatible with de Broglie's theory, and it could be easily applied to many different physical situations. This heuristic value of Schrödinger's wave equation is another very important distinction between the two theories, since de

MARTINS, Roberto de Andrade. *Studies in History and Philosophy of Science II*. Extrema: Quamcumque Editum, 2021.

Broglie's theory only led to a single new prediction: the wave behaviour of electrons in diffraction experiments.

**Keywords**: Schrödinger, Erwin; de Broglie, Louis; wave mechanics; wave equation; quantum mechanics; history of physics

## 1. INTRODUCTION

The researches of Erwin Rudolf Josef Alexander Schrödinger (1887-1961) on wave mechanics started in late 1925 as a development of his study of the 1924 thesis of Louis-Victor-Pierre-Raymond de Broglie (1892-1987). It is well known that Schrödinger's wave equation can be derived from De Broglie's results, in the classical limit. From this point of view, one might think that Schrödinger's theory seems a mere development of De Broglie's theory. However, can we really accept that conclusion?

This paper will compare some features of De Broglie's and Schrödinger's theories. It is well known that in his initial attempts to formulate a quantum theory of the atom Schrödinger tried to develop a relativistic theory, following de Broglie's ideas, and only afterwards he looked for a non-relativistic wave equation. It is straightforward to derive a wave equation for de Broglie's phase waves both in the relativistic and non-relativistic realms. In the case of his relativistic attempt, Schrödinger did indeed follow a simple approach, using de Broglie's theory. In developing the non-relativistic approach, however, Schrödinger attempted to produce an independent derivation of the wave equation, following several different lines, instead of using de Broglie's results in the classical limit.

I will first discuss the historical influence of De Broglie's work on Schrödinger; then, the early derivations of the wave equation, stressing the differences and similarities between the two theories. It will be shown that, although it is formally possible to derive a wave equation from de Broglie's theory, there is an incompatibility between the two theories: it would be impossible to make any sense of de Broglie's ideas in the case of the rigid rotator, for instance. Schrödinger's approach was, in this sense, independent and incompatible with de Broglie's theory, and it could be easily applied to many different physical situations.[1]

## 2.  DE BROGLIE'S THEORY

Louis de Broglie's theory was first presented in a series of papers published in 1923-1924 (Broglie, 1923a, 1923b, 1923c, 1923d; 1924a, 1924b, 1924c) and in his PhD thesis (Broglie, 1924d; 1925). He took as a starting point the idea that all particles (electrons, light quanta, etc.) underwent some periodical process obeying both the relativistic and quantum energy equations $E=h\nu$ and $E=mc^2$, and used special relativity as the main theoretical tool of his work (Broglie, 1923a).[2]

In the rest frame of the particle, one should have $E_0=m_0c^2=h\nu_0$ and relative to other reference systems, the correct equation should be:

$$E=mc^2=h\nu \qquad\qquad (1)$$

However, mass increases with speed, and frequency decreases with speed. Therefore, it seemed that the use of $E=mc^2=h\nu$ would lead to a contradiction (Broglie, 1923a, pp. 507-508). After dealing with this difficulty for a while, De Broglie recognised that his theory could only obey the special theory of relativity if he conceived all quanta as extended systems, instead of point particles.

---

[1] This work was written for presentation at the workshop "Quantum theory: historical studies and cultural implications", held at the Universidade Federal da Paraíba, Campina Grande, Brazil, 15-17 December 2008. A shorter Portuguese version has already been published (Martins, 2010), but this English version is now published for the first time.

[2] A detailed discussion of De Broglie's work can be found in Martins & Rosa, 2014 (in Portuguese).

In his thesis he presented this fundamental idea in a very clear way. According to Maxwell's electromagnetic theory, the energy of any charge (including an electron) is spread in the space around it, although it has a strong energy concentration around a centre. Following this idea, De Broglie regarded the electron as an infinite system (Broglie, 1925, pp. 33-34).

In the rest frame of an electron, its whole (infinite) structure was supposed to be pulsating in synchrony, with a frequency given by $h\nu_0=m_0c^2$. This periodic phenomenon, independent of space, could be described by an equation such as:

$$\Psi_0 = A \operatorname{sen} 2\pi\left(\nu_0 t_0\right) \tag{2}$$

Relative to other reference systems, the synchrony of this periodic phenomenon would be lost, of course, due to relativistic effects. The Lorentz transformation of time is:

$$t_0 = \frac{1}{\sqrt{1-\beta^2}}\left(t - \frac{\beta x}{c}\right) \tag{3}$$

where $\beta=v/c$ is the speed of the particle divided by the speed of light.

Applying the Lorentz transformation of time to this pulsation, de Broglie easily showed that the oscillation would transform to a wave, relative to other reference frames, and obtained the speed, frequency and other properties of the wave (Broglie, 1925, pp. 35-36). Replacing $t_0$ in (2) by (3), we get:

$$\Psi_0 = A \operatorname{sen} 2\pi\left[\nu_0 \frac{1}{\sqrt{1-\beta^2}}\left(t - \frac{\beta x}{c}\right)\right] \tag{4}$$

The general formula for a monochromatic wave travelling in the $x$ direction is:

$$\Psi_0 = A \operatorname{sen} 2\pi\left[\nu\left(t - \frac{x}{V}\right)\right] \tag{5}$$

Comparison of equations (4) and (5) shows that the uniform pulsation of the electron (in its proper reference frame) becomes a monochromatic wave (the "phase wave") relative to other reference systems; and by identifying the corresponding quantities in both equations, one obtains the frequency $\nu$ and speed $V$ of the wave associated to the electron:

$$\nu = \frac{\nu_0}{\sqrt{1-\beta^2}} \tag{6}$$

$$V = \frac{c}{\beta} = \frac{c^2}{\nu} \tag{7}$$

The electron should have some definite position, of course. Therefore, the uniform infinite wave cannot describe it completely. The moving free electron would be equivalent to an extended system with strong energy concentration around a centre, travelling at a speed $\nu$, and at the same time traversed by a monochromatic wave of speed $V=c^2/\nu$ and frequency $\nu=mc^2/h$.

A modulated monochromatic wave is mathematically equivalent to a wave group, but conceptually it is quite different, because in the rest frame it does have a single, well-defined frequency, and it shouldn't spread as it moves.

## 3. MECHANICS AND OPTICS

De Broglie presented his theory in several different ways. In some of his publications he emphasised the similarities between mechanics and optics (Broglie, 1924a; 1925, pp. 46-53). In the special theory of relativity, the Maupertuis' principle of least action can be written as:

$$\delta \int_{P}^{Q} J_i dx^i = 0 \tag{8}$$

where $J_i$ are the components of the energy-momentum four-vector.

On the other hand, the relativistic version of Fermat's principle can be written as:

$$\delta \int_P^Q O_i dx^i = 0 \qquad (9)$$

where $O_i$ are the components of the four-vector "universe wave", with components corresponding to the wave number projections and the frequency of the wave.

The analogy between the two principles (Fermat and Maupertuis) and the relation $E=h\nu$ then allowed De Broglie to establish a general relation in four dimensions:

$$O_i = \frac{1}{h} J_i \qquad (10)$$

which contain both $E=h\nu$ and $p=h/\lambda$ as special cases of the relativistic equation (Brown & Martins, 1984).

## 4.  ELECTROMAGNETIC FIELDS

If the electron is moving in an electromagnetic field, its total energy (including the potential energy $e\varphi$) remains constant (Broglie, 1925, p. 60). De Broglie supposed that the frequency of the electron wave should be proportional to the total energy $W$, and therefore it would also be constant.

$$h\nu = W = \frac{m_0 c^2}{\sqrt{1-\beta^2}} + e\varphi \qquad (11)$$

However, the speed V of the waves and its wavelength would change from place to place, according to a very complex equation (Broglie, 1925, p. 60):

$$V = \frac{W}{p} = \frac{\dfrac{m_0 c^2}{\sqrt{1-\beta^2}} + e\varphi}{\dfrac{m_0 \beta c}{\sqrt{1-\beta^2}} + eA_l} = \frac{c}{\beta}\frac{W}{W - e\varphi}\cdot\frac{1}{1 + eA_l/G} \qquad (12)$$

In this equation, the momentum of the electron contains components proportional to the potential vector *A*, following Maxwell's theory (Bork, 1967).

De Broglie did not attempt to apply (11) and (12) to any specific situation. The only case of a bound particle he was able to deal with was the hydrogen atom (Broglie, 1923a, pp. 509-510). He supposed that the centre of the electron obeyed classical mechanics and followed a Kepler path. He assumed that the wave would follow the same classical trajectory. Assuming that the wave should always be in phase with the electron oscillations (and not assuming that the wave should be stationary, as presented in textbooks), he proved that Bohr's quantum rule for the angular momentum $L=pr=nh/2\pi$ was a consequence of his own theory (Broglie, 1925, pp. 62-65).

The only new prediction of De Broglie's theory was electron diffraction (Broglie, 1923b, p. 549; 1925, p. 104), and this was soon confirmed. Experiments with high-energy electrons proved, in a few years, that the wavelength of the electron wave obeyed a relativistic equation, as predicted by De Broglie.

## 5. THE EINSTEIN CONNECTION

It is usually said that De Broglie's thesis was only accepted due to Albert Einstein's influence upon Paul Langevin (Mehra & Rechenberg, 1982, vol. 1.2, p. 604). This version, grounded upon De Broglie's testimony, is not correct.

Paul Langevin told Einstein about De Broglie's work in July 1924, and on July 27 he asked De Broglie to send a copy of his thesis (before it was approved) to Einstein (Wheaton, 1983, p. 297; Darrigol, 1993, p. 355). However, there was no immediate reaction from Einstein. De Broglie's thesis was presented and approved on November 25. Only on December 16 Einstein wrote letters to Langevin and to Lorentz praising De Broglie's work – "He has lifted a corner of the big veil" (Darrigol, 1993, p. 355; Mehra & Rechenberg, 1982, vol. 1.2, p. 604). On January 13 Langevin wrote a letter to De Broglie, telling him about Einstein's favourable opinion (Wheaton, 1983, p. 297).

At this time Einstein was working on the quantum theory of gases (now called "Bose-Einstein statistics"). In a paper published in February 1925 he remarked that De Broglie's work might help to elucidate the meaning of the new theory (Jammer, 1966, p. 249).

Erwin Schrödinger read Einstein's papers, and they exchanged letters about the quantum theory of gases (Hanle, 1977; 1979). Stimulated by Einstein's reference to De Broglie's work, Schrödinger obtained a copy of the thesis and read it in October 1925[3]. In November 1925 Schrödinger wrote letters to Einstein and to Landé showing that he was very excited with De Broglie's ideas (Moore, 1989, p. 192). He applied De Broglie's theory to gases in a paper he finished in December 1925 (Hegt, 1997, p. 474). However, he found some features of De Broglie's theory difficult to understand or to accept – especially the theory of the hydrogen atom.

On November 23, 1925, Schrödinger presented a seminar on De Broglie's ideas (Moore, 1989, p. 192). At that occasion, Peter Debye remarked that De Broglie's approach was childish and that it was necessary to use a wave equation to describe the wave in three dimensions (Kragh, 1982, p. 157). Schrödinger agreed that the waves should be dealt with in another way, in the case of the hydrogen atom. He also noticed that the waves

---

[3] According to Heitler (1961, p. 222) many other physicists studied De Broglie for the same reason, but nobody – except Schrödinger – took the idea of waves associated to electrons seriously. See Raman & Forman's analysis of Schrödinger's peculiar attitude towards De Broglie's work (Raman & Forman, 1969).

in nearby Keplerian orbits would produce a distorted wave front, therefore De Broglie's approach was too simplistic.

In December 1925 Schrödinger began his attempts to produce a wave equation from De Broglie's theory and to apply it to the hydrogen atom. Instead of waves following Keplerian orbits he began to think about standing waves in three dimensions, analogous to sound waves in cavities. Quantization should arise as a consequence of the discrete spectrum of standing waves in the atom.

## 6. SCHRÖDINGER'S RELATIVISTIC WAVE EQUATION

Some decisive steps were made around Christmas, during Schrödinger's stay in Villa Herwig, in the Alps, where he spent two weeks with a mysterious lover (Moore, 1989, pp. 194-195). Schrödinger first tried to produce a relativistic wave equation, following De Broglie's approach (Kragh, 1982, pp. 175-178). This derivation was not published but it can be found in a manuscript, probably written in late December 1925 (Mehra & Rechenberg, 2001, vol. 5.1, pp. 423-430). Let us present a reconstruction of Schrödinger's first derivation of the wave equation (Kragh, 1982, p. 180; Kragh, 1984).

The general wave equation, valid both in classical and relativistic physics, is:

$$\Delta\psi + \left(\frac{2\pi}{\lambda}\right)^2 \psi = 0 \tag{13}$$

For any monochromatic wave $\lambda = V/\nu$, therefore the general wave equation can also be written as a function of the speed of the wave and its frequency:

$$V^2\Delta\psi + 4\pi^2\nu^2\psi = 0 \tag{14}$$

In De Broglie's theory, $V = E/p$ where $E$ is the total energy of the electron:

$$E = h\nu = mc^2 - e\phi = (m_0c^2)/(1-\beta^2)^{1/2} - e\phi \qquad (15)$$

If magnetic fields are disregarded, the momentum $p$ is:

$$p=mv=(m_0\beta c)/(1-\beta^2)^{1/2} \qquad (16)$$

Since the total energy of the electron is equal to $h\nu$ (eq. 15,) it is possible to obtain $v=\beta c$ as a function of $\nu$. Substituting this result in the equations for E and p one can compute the wave velocity $V = E/p$ as a function of the frequency $\nu$. After some manipulation, the general wave equation (13) becomes:

$$\frac{h^2c^2}{4\pi^2}\Delta\psi + \left[\left(E - e\varphi\right)^2 - m_0^2c^4\right]\psi = 0 \qquad (17)$$

This is the so-called "Klein-Gordon equation".

Notice that Schrödinger's derivation of the relativistic wave equation depends only on results that had already been obtained by De Broglie. Indeed, De Broglie himself did also arrive to the same result, independently (Kragh, 1984, p. 1025).

Schrödinger applied this wave equation to the hydrogen atom and obtained *wrong* results for the energy levels (Jammer, 1966, pp. 257-258; Mehra & Rechenberg, 2001, vol. 5.1, pp. 367-368). After struggling for a short time with the relativistic theory he turned to a non-relativistic approach.

## 7. THE CLASSICAL WAVE EQUATION

Obtaining a wave equation in the classical approximation is much easier than in the relativistic case, and many textbooks present such a derivation. If one accepts the relation $\lambda=h/p$ between momentum and wavelength and applies classical dynamics, one obtains:

$$\lambda = \frac{h}{p} = \frac{h}{mv} \qquad (18)$$

In classical mechanics, the kinetic energy $K$ is:

$$K = \frac{m\mathrm{v}^2}{2} = E - U \qquad (19)$$

Therefore, the square of wavelength, in the classical limit, is:

$$\lambda^2 = \frac{h^2}{p^2} = \frac{h^2}{m^2\mathrm{v}^2} = \frac{h^2}{2m(E-U)} \qquad (20)$$

The general (classical and relativistic) wave equation is:

$$\Delta\psi + \left(\frac{2\pi}{\lambda}\right)^2 \psi = 0 \qquad (21)$$

Replacing $\lambda^2$ in (21) by the expression in (20), one obtains:

$$\Delta\Psi + \frac{8\pi^2 m}{h^2}(E-U)\Psi = 0 \qquad (22)$$

This is the so-called Schrödinger equation independent of time. Therefore, using only classical mechanics and De Broglie's relation $\lambda = h/p$ it is possible to derive the Schrödinger wave equation. Notice that this derivation is very similar to Schrödinger's relativistic derivation shown above, but much simpler. However, Schrödinger did not use this very simple derivation. How did Schrödinger present the wave equation in his early papers?[4]

## 8. THE WAVE EQUATION IN SCHRÖDINGER'S FIRST PAPER

In his first 1926 paper Schrödinger presented the wave equation as a consequence of the Hamilton-Jacobi approach, in a very abstract way (Schrödinger, 1926a, pp. 361-362; 1929, pp.

---

[4] I will not discuss here how Schrödinger arrived to his equation. The published papers do not present his ideas as they were really developed (Kragh, 1982, p. 158). What interests me is how Schrödinger chose to publish his results and what could be his motivation for presenting them exactly in this way.

1-2). He introduced an unknown function $\Psi$ and stated that the action $S$ could be written as

$$S = K \log \Psi \qquad (23)$$

The Hamiltonian function $H$ could therefore be written as:

$$H\left(q, \frac{\partial S}{\partial q}\right) = H\left(q, \frac{K}{\Psi} \frac{\partial \Psi}{\partial q}\right) = E \qquad (24)$$

Schrödinger then stated that in the non-relativistic case, this equation "can always be transformed so as to become a quadratic form (of $\Psi$ and its derivatives) equated to zero" (Schrödinger, 1926a, pp. 361-362; 1929, p. 1), and in the case of the hydrogen atom (a Coulombian field) this would become:

$$\left(\frac{\partial \Psi}{\partial x}\right)^2 + \left(\frac{\partial \Psi}{\partial y}\right)^2 + \left(\frac{\partial \Psi}{\partial z}\right)^2 - \frac{2m}{K^2}\left(E + \frac{e^2}{r}\right)\Psi^2 = 0 \qquad (25)$$

This is just the classical equation of energy conservation $p^2=2m(E-V)$, since the three first terms correspond the squared momentum divided by $(K/\Psi)$, according to (24).

Schrödinger than stated (Schrödinger, 1926a, p. 362; 1929, p. 2): "We now seek a function $\Psi$ such that for any arbitrary variation of it the integral of the said quadratic form, taken over the whole co-ordinate space, is stationary [...]". The corresponding equation is:

$$\delta J = \delta \left[\left(\frac{\partial \Psi}{\partial x}\right)^2 + \left(\frac{\partial \Psi}{\partial y}\right)^2 + \left(\frac{\partial \Psi}{\partial z}\right)^2 \right.$$
$$\left. - \frac{2m}{K^2}\left(E + \frac{e^2}{r}\right)\Psi^2 \right] dx.\,dy.\,dz = 0 \qquad (26)$$

From this variational problem Schrödinger derived the wave equation for the hydrogen atom:

$$\nabla^2 \Psi + \frac{2m}{K^2}\left(E + \frac{e^2}{r}\right)\Psi = 0 \qquad (27)$$

This derivation presented in Schrödinger's first 1926 paper is completely meaningless, since $\Psi$ was an undefined function and (26) is not a valid variational principle in classical physics[5]. Its only "justification" is that the wave equation leads to the correct energy levels of the hydrogen atom, carefully derived in the rest of the same paper (Schrödinger, 1926a, p. 362-374; 1929, pp. 2-10).

## 9. THE DERIVATION IN SCHRÖDINGER'S SECOND PAPER

Schrödinger himself was not satisfied with this "derivation", and presented a very different one in his second 1926 paper (Schrödinger, 1926b; Schrödinger, 1929, pp. 13-40), where he presented for the first time the time-independent wave equation ($m=1$)

$$\Delta\Psi + \frac{8\pi^2 m}{h^2}(E - U)\Psi = 0 \qquad (28)$$

We present here a reconstruction of his derivation, stressing only its main points.

Schrödinger's fundamental assumption was the general (classical) wave equation in this form (Schrödinger, 1926b, p. 510; Schrödinger, 1929, p. 27)[6]:

---

[5] "The first published derivation [...] was not only curiously formal, but straighforwardly cryptical. On the whole this derivation appears badly justified, its sole foundation lying in its result [...]" (Kragh, 1982, p. 158).

[6] Instead of the usual symbol for the Laplacian operator used here, Schrödinger wrote *div grad*.

$$\nabla^2\psi - \frac{1}{u^2}\ddot{\psi} = 0 \qquad (29)$$

Schrödinger had already obtained (Schrödinger, 1926b, pp. 494-498; Schrödinger, 1929, pp. 16-20) the equation of the speed of the wave ($m=1$):

$$u = \frac{ds}{dt} = \frac{E}{\sqrt{2(E-V)}} = \frac{h\nu}{\sqrt{2(h\nu - V)}} \qquad (30)$$

Now, assuming that the wave function $\Psi$ has the same form it usually has in classical physics (Schrödinger, 1926b, p. 510; Schrödinger, 1929, p. 27):

$$\psi = f(q_k).e^{2\pi i \nu t} \qquad (31)$$

we derive:

$$\ddot{\psi} = \frac{\partial^2\psi}{\partial t^2} = (2\pi i \nu)^2.f(q_k).e^{2\pi i \nu t} = -4\pi^2\nu^2\psi \qquad (32)$$

By replacing (32) and (30) in (29) one obtains at once Schrödinger's wave equation ($m=1$) :

$$\nabla^2\psi + \frac{8\pi^2}{h^2}(h\nu - V)\psi = 0 \qquad (33)$$

Since $E=h\nu$, it may also be written as (Schrödinger, 1926b, p. 510; Schrödinger, 1929, p. 27):

$$\nabla^2\psi + \frac{8\pi^2}{h^2}(E - V)\psi = 0 \qquad (34)$$

Therefore, in this derivation presented in Schrödinger's second 1926 paper, the only non-classical assumptions are $E=h\nu$ and the formula for the speed $u$ of the waves associated to the electron (30) that corresponds to the classical limit of De Broglie's relation $u=E/p$.

However, Schrödinger did not take the formula for $u$ from De Broglie's work. He presented his own, original and highly abstract derivation of this relation, using the analogy between the principles of Huygens and Hamilton (Schrödinger, 1926b, pp. 494-498; Schrödinger, 1929, pp. 16-20). He referred to De Broglie's work in this paper, but stressed that his own results were obtained in a more general way, and independently of the theory of relativity. After presenting his own derivation, he remarked:

> We find here again a theorem for the 'phase waves' of the electron, which Mr. de Broglie had derived, with essential reference to the relativity theory, in those fine researches to which I owe the inspiration for this work. We see that the theorem in question is of wide generality, and does not arise solely from relativity theory, but is valid for every conservative system of ordinary mechanics. (Schrödinger, 1926b, p. 498; Schrödinger, 1929, p. 20)

So, Schrödinger *could* have used De Broglie's work to derive the wave equation in the non-relativistic case, *but he did not do that*. He used a formula for the speed of the wave associated to the electron that is equivalent to De Broglie's, but he provided a new derivation of this relation – seemingly because he wanted to prove that it was valid independently of relativistic considerations.

## 10. SCHRÖDINGER'S AND DE BROGLIE'S THEORIES

We might justify Schrödinger's careful avoidance of grounding his own work upon De Broglie's theory in the following way. He had first attempted to use the relativistic approach, and it failed; besides that, the wave equation that did lead to correct results is not the classical limit of the relativistic wave equation. Therefore, it was appropriate to present a derivation completely independent of De Broglie's relativistic theory, to ensure that his own theory was well grounded. Probably this was part of his motivation. However, there are other relevant issues, since there are several deep differences between the two theories.

De Broglie's work was grounded upon special relativity, and its main equations could only be deduced for electrons in uniform motion, since he took as his starting point the description of the pulsation of the extended electron in its rest frame. Its extension to accelerated motion required a new way of thinking, and a different justification. De Broglie did attempt to provide a basis for his theory in the case of accelerated motion using the analogy between the principles of Maupertuis and Fermat – that are valid only for single particles and wave rays. This led him to a general relation between the wave properties and the mechanical properties of the electron moving in an electromagnetic field. However, De Broglie's electron still described a definite trajectory and the associated wave could be described by wave rays – an approximation which is adequate in describing phenomena corresponding to geometrical optics, but completely inadequate for the analysis of diffraction and interference. He successfully applied his ideas to the hydrogen atom, but he was unable to study the harmonic oscillator and other simple systems. De Broglie's concept of the electron could not be adequately applied when the dimensions of the system was comparable to the wavelength (as in the case of the atom). It was also difficult to perceive how his theory could be applied to a system with many particles. Besides that, his theory would be meaningless for a rigid rotator, for instance.[7]

Schrödinger's approach was much more general than De Broglie's. His derivations did not depend on the theory of relativity and therefore he could directly address the case of accelerated and rotating systems. Besides that, in the derivation presented in his second 1926 paper (Schrödinger, 1926b, 490-491; 1928, p. 14) he presented his own theory in a very general

---

[7] The discussion of the quantum rigid rotator was very important in the theory of specific heat of poliatomic gases. This problem was addressed by Schrödinger in his second 1926 paper.

way, using general coordinates (therefore opening the possibility of applying it to rotation and other types of motion) and discussing a general conservative system (that is, he did not restrict his treatment to a single particle). His general and abstract way of approaching the problem allowed him to apply the wave equation to any physical system.

In his second paper, Schrödinger used the analogy between the principles of Huygens and Hamilton (not Fermat and Maupertuis, as De Broglie did), and that step ensured that the relation could be carried to cases where the dimensions of the system were comparable to the wavelength.

For those reasons, only Schrödinger's theory could be directly applied to waves associated to one or more electrons in three dimensions (atomic systems), to the harmonic oscillator and to rotating solids, as he did (Schrödinger, 1926b).

## 11. CONCLUSIONS

Schrödinger's wave equation can be derived from De Broglie's results, in the classical limit. However, Schrödinger's theory is not an application or development of De Broglie's theory.

Schrödinger's theory can be applied to cases where De Broglie's theory cannot be applied, such as accelerated motion and rotation, and for bound particles where the wavelength is comparable to the dimensions of the region containing the electron. Even in the case of a free particle they are also incompatible: De Broglie's *relativistic* wavelength was confirmed by diffraction experiments, and Schrödinger's classical wavelength is simply wrong, for high speeds.

The heuristic value of Schrödinger's wave equation is another very important distinction between the two theories, since de Broglie's theory only led to a single new prediction: the wave behaviour of electrons in diffraction experiments. There are also other differences between the two approaches that cannot be described here.

Although De Broglie's theory was the starting point of Schrödinger's work and had a very important heuristic role in this respect, the two theories are different, independent and incompatible.

## ACKNOWLEDGEMENTS

The author is grateful to the Brazilian National Council for Scientific and Technological Development (CNPq) and to the São Paulo State Research Foundation (FAPESP) for supporting this research.

## BIBLIOGRAPHIC REFERENCES

BORK, Alfred M. Maxwell and the vector potential. *Isis*, **58**: 210-222, 1967.

BROGLIE, Louis de. Ondes et quanta. *Comptes Rendus de l'Académie des Sciences de Paris*, **177**: 507-510, 1923 (a).

BROGLIE, Louis de. Quanta de lumière, diffraction et interférences. *Comptes Rendus de l'Académie des Sciences de Paris*, **177**: 548-550, 1923 (b).

BROGLIE, Louis de. Les quanta, la théorie cinétique des gaz et le principe de Fermat. *Comptes Rendus de l'Académie des Sciences de Paris*, **177**: 630-632, 1923 (c).

BROGLIE, Louis de. Waves and quanta. *Nature*, **112**: 540, 1923 (d).

BROGLIE, Louis de. A tentative theory of light quanta. *Philosophical Magazine*, **47**: 446-458, 1924 (a).

BROGLIE, Louis de. Sur la définition générale de la correspondance entre onde et mouvement. *Comptes Rendus de l'Académie des Sciences de Paris*, **179**: 39-40, 1924 (b).

BROGLIE, Louis de. Sur un théorème de M. Bohr. *Comptes Rendus de l'Académie des Sciences de Paris*, **179**: 676-677, 1924 (c).

BROGLIE, Louis de. *Thèses. Recherches sur la théorie des quanta.* Paris: Masson, 1924 (d).

BROGLIE, Louis de. Recherches sur la théorie des quanta. *Annales de Physique*, series 10, **3**: 22-128, 1925.

BROWN, Harvey Robert; MARTINS, Roberto de Andrade. De Broglie's relativistic phase waves and waves groups. *American Journal of Physics*, **52**: 1130-1140, 1984.

DARRIGOL, Olivier. Strangeness and soundness in Louis de Broglie's early works. *Physis: Rivista Internazionale di Storia della Scienza*, **30**: 303-372, 1993.

HANLE, Paul A. Erwin Schrödinger's reaction to Louis de Broglie's thesis on the quantum theory. *Isis*, **68**: 606-609, 1977.

HANLE, Paul A. The Schrödinger-Einstein correspondence and the sources of wave mechanics. *American Journal of Physics*, **47**: 644-648, 1979.

HEITLER, Walter. Erwin Schrödinger, 1887-1961. *Biographical Memoirs of Fellows of the Royal Society*, **7**: 221-228, 1961.

JAMMER, Max. *The conceptual development of quantum mechanics*. New York: MacGraw-Hill, 1966.

KRAGH, Helge. Erwin Schrödinger and the wave equation: the crucial phase. *Centaurus*, **26**, 154-197, 1982.

KRAGH, Helge. Equation with the many fathers. The Klein-Gordon equation in 1926. *American Journal of Physics*, **52**, 1024-1033, 1984.

MARTINS, Roberto de Andrade. De Louis de Broglie a Erwin Schrödinger: uma comparação. Pp. 393-409, in: FREIRE JR, Olival; PESSOA JR., Osvaldo; BROMBERG, Joan Lisa (orgs). *Teoria quântica: estudos históricos e implicações culturais*. Campina Grande: EDUEPB; São Paulo: Livraria da Física, 2010.

MARTINS, Roberto de Andrade; ROSA, Pedro Sérgio. *História da teoria quântica: a dualidade onda-partícula, de Einstein a De Broglie*. São Paulo: Livraria da Física, 2014.

MEHRA, Jagdish; RECHENBERG, Helmut. *The historical development of quantum theory*. 5 vols. New York: Springer, 1982-1987.

MOORE, W. J. *Schrödinger: life and thought*. Cambridge: Cambridge University, 1989.

RAMAN, V. V.; FORMAN, Paul. Why was it Schrödinger who developed de Broglie's ideas? *Historical Studies in the Physical Sciences*, **1**: 291-314, 1969.

REGT, Henk W. de. Erwin Schrödinger, Anschaulichkeit, and quantum theory. *Studies in History and Philosophy of Modern Physics*, **28**: 461-481, 1997.

SCHRÖDINGER, Erwin. Quantisierung als Eigenwertproblem (Erste Mitteilung). *Annalen der Physik*, series 4, **79**: 361-376, 1926 (a).

SCHRÖDINGER, Erwin. Quantisierung als Eigenwertproblem (Zweite Mitteilung). *Annalen der Physik*, series 4, **79**: 489-527, 1926 (b).

SCHRÖDINGER, Erwin. *Collected papers on wave mechanics*. London & Glasgow: Blackie & Son, 1929.

WHEATON, Bruce R. *The tiger and the shark. Empirical roots of wave-particle dualism*. London: Cambridge University, 1983.

# THE CULTURAL RELEVANCE OF ASTRONOMY IN CLASSICAL ANTIQUITY

Roberto de Andrade Martins

**Abstract**: Nowadays we understand 'astronomy' as the study of celestial bodies. However, astronomical knowledge in classical Greece and Rome cannot be reduced to the study of the sky. Of course, it did include the study of stars and planets, but this knowledge was intimately linked to many other subjects, such as religion, geography, meteorology, mythology, medicine, etc. Acquaintance with many celestial phenomena and the constellations and their lore was part of ancient culture. Remove from the classical Antiquity its astronomical knowledge, and many of its features and activities would become impossible or meaningless. Astronomy was deeply implanted in ancient culture, as will be shown in this paper, and that was the reason underlying its very high status in Antiquity.
**Keywords**: Greek astronomy; Roman astronomy; astrology; cultural astronomy; history of astronomy

## 1.  INTRODUCTION

Nowadays, astronomy is a serious concern solely for astronomers. A few non-astronomers have an intellectual curiosity for astronomical knowledge, but modern astronomy has no direct influence on the life of most people. If astronomical knowledge just disappeared tomorrow, this would have negligible impact on everyday activities. The situation was completely different, however, in the Greek and Roman worlds.

MARTINS, Roberto de Andrade. *Studies in History and Philosophy of Science II*. Extrema: Quamcumque Editum, 2021.

This paper will show how deep was the influence of celestial knowledge in many fields of ancient culture – agriculture, navigation, religion, geography, medicine, meteorology, astrology, time measurement and other subjects.[1] Acquaintance with many celestial phenomena and the constellations and their lore was part of ancient culture. Remove from the classical Antiquity its astronomical knowledge, and many of its features and activities would become impossible or meaningless.

For understanding the cultural relevance of astronomy in the classical culture it is necessary to realize that its focus was not planetary theory, as one might think after consulting a few works on the history of astronomy. We are the heirs of the "Copernican revolution" and for that reason we usually think that the main astronomical question was whether the Earth is the unmoving center of the universe or an ordinary planet moving around the Sun. All extant histories of astronomy were written after Copernicus, and they have an unquestionable Whig style. Copernicus' contribution was a change in the understanding of the motion of the planets, and introduced the motion of the Earth itself; therefore – that is the Whig catch – the history of astronomy should be the account of the ideas about those two subjects. However, the study of planetary theory was just a small part of astronomy, before the Copernican revolution; and the very meaning of the old astronomical culture is absent in most recent accounts.

The astronomical status of the Earth and the motion of the planets was not the heart of Greek and Roman astronomy. Its core was the study of the stars and constellations, of the Sun, and the Moon. At a time when people spent much time in open spaces and artificial lighting was feeble, knowledge of the stars

---

[1] This paper and its sequel (the next essay in this volume) were written for presentation at the conference *Oxford Scientiae 2016: Disciplines of knowing in the early modern world* – St Anne's College, University of Oxford, 5-7 July 2016. It was circulated at that time, but it had not been published until now.

and the constellations, the apparent motion of the Sun and the study of the shadows it produces, the variable duration of the day and night during the year, the phases of the Moon and other conspicuous phenomena were well known. The motion of the planets, on the other hand, was not described in the oldest extant Greek astronomical works. Notice that the Moon moves around the Earth both in geocentric and heliocentric astronomies; and the apparent motion of the stars and of the Sun is much easier to understand from the point of view of a geocentric theory, which can deal with all its phenomena.

However, astronomy in Antiquity cannot be reduced to the study of the sky, because of its intimate relationship with many other subjects, such as religion, geography, meteorology, mythology, medicine, etc. Astronomy was deeply implanted in ancient culture, as will be shown in this paper. That was the reason underlying its very high status in Antiquity – and that was lost during the early modern period, as will be pointed out in a next paper.[2]

## 2. THE ROLES OF ASTRONOMY IN CLASSICAL ANTIQUITY

Since Antiquity, astronomy was regarded as a central component of the human culture[3]. In Aeschylus' *Prometheus Bound*, the main character thus describes the very birth of mankind:

> First of all, though they had eyes to see, they saw to no avail; they had ears, but they did not understand; but, just as shapes in dreams, throughout their length of days, without purpose they wrought all things in confusion. They had

---

[2] MARTINS, Roberto de Andrade. The transformation of astronomical culture in the seventeenth century, in this volume.

[3] This paper will only address the European case. Of course, in other regions of the world, astronomy was also an integral part of culture, since Antiquity or Pre-History. See, for instance, Kelley & Milone (2011) for a survey of early astronomy in other cultures.

neither knowledge of houses built of bricks and turned to face the sun nor yet of work in wood; but dwelt beneath the ground like swarming ants, in sunless caves. They had no sign either of winter or of flowery spring or of fruitful summer, on which they could depend but managed everything without judgment, until I taught them to discern the risings of the stars and their settings, which are difficult to distinguish. Yes, and numbers, too, chiefest of sciences, I invented for them, and the combining of letters, creative mother of the Muses' arts, with which to hold all things in memory. I, too, first brought brute beasts beneath the yoke to be subject to the collar and the pack-saddle, so that they might bear in men's stead their heaviest burdens; and to the chariot I harnessed horses and made them obedient to the rein, to be an image of wealth and luxury. It was I and no one else who invented the mariner's flaxen-winged car that roams the sea. (Aeschylus, *Prometheus Bound*, 447-468)

In Aeschylus' account, the basic astronomical knowledge of the stars, their risings and settings, and the signs of the seasons, together with mathematics, writing and the basic techniques, were taught by Prometheus to the brutes who became the early human tribe.

## 3. HESIOD'S YEAR

Some of the relevant traits of astronomy in the early Greek culture appear in Hesiod's *Works and Days* (Ἔργα καὶ Ἡμέραι). In this poem, the year – an astronomical concept – and its seasons and phenomena are described by the appearance of specific stars or constellations, and by solar events, such as the solstices. The knowledge of those phenomena was essential, in agriculture.

When the Pleiades, daughters of Atlas, are rising [late in May], begin your harvest, and your ploughing when they are going to set [in November]. Forty nights and days they are hidden and appear again as the year moves round, when first you sharpen your sickle. (Hesiod, *Works and Days*, 383-387)

Notice that the Pleiades were described as the daughters of Atlas; they were not just a set of stars; they had a mythological meaning – as any other astronomical entity. A thorough acquaintance with the sky required the knowledge of mythology, and vice-versa. The Greeks identified stars with heroes and heroines who were translated to the sky by the gods. A lost work attributed to Hesiod, with the title *Astronomia*, described the myths associated to the stars and other heavenly bodies. The earliest known work on this subject was written by Eratosthenes, in the third century BCE (Davidson, 2007, p. 206).

The very concept of time and its divisions was a consequence of the study of the sky. The division of the year among the Greek obeyed the phenomena of two different astronomical bodies: the Sun and the Moon. The months began at new Moon, usually regarded as the first appearance of the lunar crescent. The mean lunar cycle takes about 29.5 days, therefore the months were either 30 or 29 days long; the year had 12, and sometimes 13 lunar months (James, 1998, p. 184). The calendar varied from city to city, beginning at different times, in each case (Gawlinski, 2016, vol. 1, pp. 891, 898). Instead of a calendar with months and days, it was more useful, in ancient Greece, to use tables of the main astronomical phenomena (appearances and disappearances of stars) that occurred in the yearly cycle. Those tables were called *parapegma*; they were, of course, the same for all Greek cities. They also included weather predictions (James, 1998, p. 190).

For Hesiod, spring begins with the late rising of Arcturus (about 24 February of our calendar), two months after the winter solstice; the beginning of winter corresponds to the setting of the Pleiades, or the early setting of the Hyades or Orion (early November). The time for harvest is given by the early rising of the Pleiades (about 19 May), threshing time by the early rise of Orion (9 July), and vintage time by the early rising of Arcturus

(18 September). He was acquainted with the solstices, but did not refer to the equinoxes (Heath, 1913, pp. 10-11).

The production of wine, the sacred drink of Dionysus, was also regulated by the stars:

> But when Orion and Sirius are come into mid-heaven, and rosy-fingered Dawn sees Arcturus [in September], then cut off all the grape-clusters, Perses, and bring them home. Show them to the Sun ten days and ten nights: then cover them over for five, and on the sixth day draw off into vessels the gifts of joyful Dionysus. But when the Pleiades and Hyades and strong Orion begin to set [the end of October], then remember to plough in season: and so, the completed year will fitly pass beneath the earth. (Hesiod, *Works and Days*, 609-617)

The adequate time for cutting wood was also related to astronomical phenomena:

> When the piercing power and sultry heat of the Sun abate, and almighty Zeus sends the autumn rains [in October], and men's flesh comes to feel far easier – for then the Dog Star [Sirius] passes over the heads of men, who are born to misery, only a little while by day and takes greater share of night – then, when it showers its leaves to the ground and stops sprouting, the wood you cut with your axe is least liable to worm. (Hesiod, *Works and Days*, 414-422)

Of course, this entails that every artisan who used wood in his work – the carpenter, architect, shipbuilders, and the persons who built furniture, wagons and other implements – all of them had to know and obey the rhythm of nature, written on the stars.

The appearance of the Dog Star (Sirius) was regarded as ominous. Euripedes' play *Iphigenia at Aulis* begins with Agamemnon's sighting of this star just before dawn, therefore associating the sacrifice of his daughter to a particularly hot and unhealthy time of the year, and to the rise of the Etesian wind (Davidson, 2007, p. 206).

Along the year, the duration of the days and nights keep varying – long nights in winter, long days in summer – and many features of everyday life should be adjusted to those changes – including the diet of men and animals:

> In this season [Winter] let your oxen have half their usual food, but let your man have more; for the long nights decrease their needs [of the animals]. Keep the habit of regulating nourishment during the whole year according to the length of days and nights, until the Earth, the mother of all, bears again her various fruit. (Hesiod, *Works and Days*, 559-563)

Not only plants, but also animals, obey the influences of the sky and appear at definite times of the year:

> When Zeus has finished sixty wintry days after the solstice [in February], then the star Arcturus leaves the holy stream of Ocean and first rises brilliant at dusk. After him the shrilly wailing daughter of Pandion, the swallow, appears to men when spring is just beginning. Before she comes, prune the vines, for it is best so. But when the house-carrier [the snail] climbs up the plants from the earth to escape the Pleiades [in the middle of May], then it is no longer the season for digging vineyards, but to whet your sickles and rouse up your slaves. (Hesiod, *Works and Days*, 564-575)
>
> But when the artichoke flowers [in June], and the chirping grass-hopper sits in a tree and pours down his shrill song continually from under his wings in the season of wearisome heat, then goats are plumpest and wine sweetest; women are most wanton, but men are feeblest, because Sirius parches head and knees and the skin is dry through heat. (Hesiod, *Works and Days*, 582-588)

Notice that the whole annual drama is directly connected with the Greek gods; and we should not regard the Earth or the Ocean as mere geographical names.

The tradition of including astronomical information in poetry was later reproduced in the Roman culture by Ovid (*Fasti*) and

Virgil (*Georgics*), for instance (James, 1998, p. 18). Columella's work on farming (1st century CE) also included detailed astronomical information.

Sailors should also be aware of the adequate winds and sailing times, according to the signs of the sky:

> But if desire for uncomfortable sea-faring seizes you; when the Pleiades plunge into the misty sea [end of October or beginning of November] to escape Orion's rude strength, then truly gales of all kinds rage. Then keep ships no longer on the sparkling sea, but bethink you to till the land as I bid you. Haul up your ship upon the land and pack it closely with stones all around to keep off the power of the winds which blow damply, and draw out the bilge-plug so that the rain of heaven may not rot it. (Hesiod, *Works and Days*, 618-626)

> Fifty days after the solstice [July-August], when the season of wearisome heat is come to an end, is the right time for me to go sailing. Then you will not wreck your ship, nor will the sea destroy the sailors, unless Poseidon the Earth-Shaker be set upon it, or Zeus, the king of the deathless gods, wish to slay them; for the issues of good and evil alike are with them. At that time the winds are steady, and the sea is harmless. (Hesiod, *Works and Days*, 663-670)

Of course, the will of Zeus, Poseidon and the other gods is unpredictable; but those who are acquainted with the language of the heaven can choose adequate times for their undertakings: "Mark the days which come from Zeus, duly telling your slaves of them, and that the thirtieth day of the month is best for one to look over the work and to deal out supplies. For these are days which come from Zeus the all-wise, when men discern aright" (Hesiod, *Works and Days*, 765-768).

The relation between the celestial phenomena and weather changes was regarded as one of cause and effect, according to Aristotle and most Greek authors, although some later authors – such as Geminus – denied this interpretation (James, 1998, p. 200).

Each day of the Moon cycle may be suitable or inadequate for different activities:

> To begin with, the first, the fourth, and the seventh [days after the New Moon] – on which Leto bare Apollo with the blade of gold – each is a holy day. The eighth and the ninth, two days at least of the waxing month, are especially good for the works of man. Also, the eleventh and twelfth are both excellent, alike for shearing sheep and for reaping the kindly fruits; but the twelfth is much better than the eleventh, for on it the airy-swinging spider spins its web in full day, and then the wise one [the ant] gathers her pile. On that day woman should set up her loom and get forward with her work. (Hesiod, *Works and Days*, 770-779)

In Hesiod's time the month was not divided in weeks, but in three parts of ten days each – the waxing, the midmonth, and the waning (James, 1998, p. 5).

## 4. RELIGIOUS FESTIVALS AND THE STARS

Time in Greece was not an abstract entity, but meant the cycles of the phenomena involving the stars, the Sun and the Moon (Davidson, 2007, p. 204). In Plato's Laws, the Athenian suggests the inclusion of astronomy in the education of the ideal city to ensure "the proper ordering of days into monthly periods, and of months into a year, so that times, sacrifices, and feasts may each be assigned their due position, according to nature [*kata physin*]" (Plato, *Laws*, 809d, *apud* Davidson, 2007, p. 206).

The Moon lies at the core of Greek calendars, resembling the calendars of Mesopotamia and Assyria; the evening sighting of the new moon's crescent was the sign of a new month (Hannah, 2005, p. 27). In the Sumerian calendars, the names of the months were derived from the agricultural activities and seasonal phenomena associated to those periods of the year. In Greece, the months were named either after gods who were honored in those months, or after associated religious festivals that took place in them. The specific names of the months varied, however, from city to city (Scullion, 2007, p. 190). There was a festival dedicated to Helios (the Sun god) in the summer solstice, and another to Cronos at the spring equinox. The Olympic festival was held every fifty moons; it began at the second full moon after the summer solstice. Many of the festivals coincided with the full moon, that is, the middle of the lunar month (Davidson, 2007, pp. 204-205). In general, it was the Moon that ruled the religious calendar and the festivals (Bouché-Leclercq, 1899, p. 45).

The lunar month was divided in three decades (an Egyptian influence); they did not use the seven-day week (Hannah, 2005, p. 87). As the time from one new moon to the following one is about 29.5 days, the division of the month in three decades was convenient: ten days of waxing, the intermediary ten days around the full moon, and ten days of waning, which were counted backwards in Athens, beginning from 9 – the numbers diminishing as the moon vanishes. There were monthly festivals that occurred at fixed days, such as Artemis' festival on day 6 (Davidson, 2007, p. 210).

The "new year" of each Greek city, associated to some relevant astronomical event (usually the solstices or equinoxes), was the time when annual magistrates or sacred officials took up office, and when new citizens were admitted. It was not just a single day, but "a much longer period, often marked by rituals of cleansing and renewal, veiling and unveiling, passing on secrets, absence and return, by festivals of disorder or of suspension of norms, and of rebirth, festivals which often looked back to the foundation of the city and/or to the start of a new divine order" (Davidson, 2007, p. 209).

The mysteries of Demeter and Kore were a central part of Greek religion. Demeter was the goddess of the harvest, who presided over grains (especially the cereals) and the fertility of the earth; she also presided over the sacred law, and the cycle of life and death. She and her daughter Kore (called Persephone in the underworld, after she was abducted by Hades) were the central figures of the Eleusinian Mysteries. Their Athens festival was held every year from the 15th to the 23rd of the month of Boedromion, roughly corresponding to September and the beginning of Autumn. In the evening of the 21st the initiates entered the sanctuary and took part in the secret rite (Clinton, 2007, p. 345).

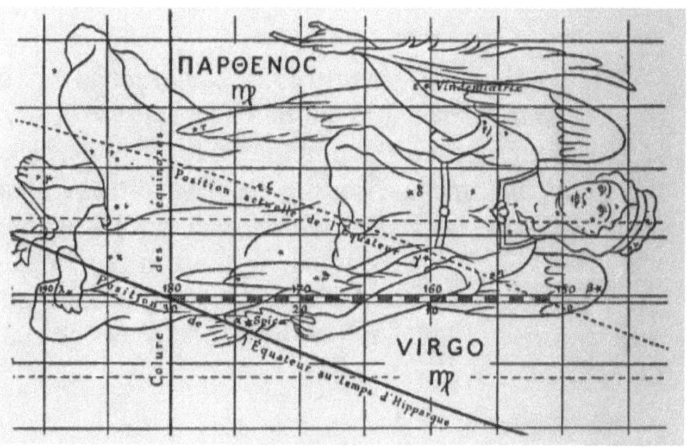


**Fig. 1.** An old representation of the Virgo constellation (Bouché-Leclercq, 1899, p. 140)

There is both a lunar and a solar symbolism behind the time of the festival. The 21st day of the lunar month was the first of the waning decade, meaning the beginning of the disappearance of the Moon. The myth of Kore was deeply related to her disappearance and the search by her mother in the underworld – with a symbolism related to the "death" of the seeds in the ground, before the crops are born. The festival was held when the Sun was in the sign of Virgo, represented by a woman holding a wheat ear (Spica) (Fig. 1); and the cereals were intimately related to Demeter (Bouché-Leclercq, 1899, p. 139; Hard, 2015, p. x). According to the Geminus' *Parapegma*, the 24th day after the Sun enters Virgo is the time when the Wheat-Ear of the Virgin rises (its morning rising); and it rains (Evans, 1998, p. 199). Of course, the solar phenomenon (the rise of Spica) will usually occur at a day different from the lunar phenomenon (the 21st day of the lunar month), but they both happen around the same period of the year.

## 5.  THE DIRECTIONS OF SPACE

Some knowledge of astronomy was also essential for sailors, because the North direction was identified by means of the stars, at a time when the magnetic compass was unknown:

> Gladly then did goodly Odysseus spread his sail to the breeze; and he sat and guided his raft skillfully with the steering-oar, nor did sleep fall upon his eyelids, as he watched the Pleiades, and late-setting Bootes, and the Bear[4], which men also call the Wain, which ever circles where it is and watches Orion, and alone has no part in the baths of Ocean. For Calypso, the beautiful goddess had bidden him to keep this star on the left hand as he sailed over the sea. (Homer, *Odyssey*, V.268-277)

Orientation, on the earth or at the sea, needed astronomical knowledge. The circumpolar stars were the best guides for finding the north, during the night. The Sun can also be used to find the four cardinal points, using a *gnomon*, but it is necessary to know that it does not rise exactly at the East and does not set exactly at the West, except for two days in each year (the equinoxes). Acquaintance with the annual motion of the Sun was necessary to understand this.

In several ancient civilizations, temples and other important buildings – such as palaces – were built following astronomical orientation. In the late 19th century, Francis Cranmer Penrose and Joseph Norman Lockyer claimed that several Greek temples exhibited relevant astronomical alignments, related to the Sun and to the rising of some stars (Kelley & Milone, 2011, p. 240). This is, however, a controversial issue. Greek temples were not erected following the four cardinal directions, and hitherto there

---

[4] During Homer's time, the name "Bear" likely included all the stars the move around the north celestial pole (Tannery, 1893, p. 7).

is no general clear interpretation of their alignment. However, in some cases it is possible that their orientation was related to specific appearances related to the Sun, the Moon and stars (Salt, 2009).

## 6. ARATOS' PHAINOMENA

The deep relation between heavenly phenomena and practical human interests is displayed not only in the early poetic productions but also in the Greek technical literature that developed from the fourth century BCE onwards. Eudoxos of Cnidos (ca. 408-355 BCE), who was a pupil of Plato and contemporaneous of Aristotle, is famous for his mathematical explanation of the motion of the planets using homocentric spheres. However, he did also compose a work on the visible celestial phenomena (now lost, as all his other books), that was rendered into verses in the third century BCE by Aratos of Solis (ca. 315-240 BCE). His poem begins by invoking Zeus:

> From Zeus let us begin; him do we mortals never leave unnamed; full of Zeus are all the streets and all the market-places of men; full is the sea and the havens thereof; always we all have need of Zeus. For we are also his offspring; and he in his kindness unto men giveth favorable signs and wakened the people to work, reminding them of livelihood. He tells what time the soil is best for the labor of the ox and for the mattock, and what time the seasons are favorable both for the planting of trees and for casting all manner of seeds. For himself it was who set the signs in heaven, and marked out the constellations, and for the year devised what stars chiefly should give to men right signs of the seasons, to the end that all things might grow unfailingly. (Aratos, *Phainomena*, 1-13)

This preamble and other parts of the poem were strongly influenced by the Stoic doctrine that the cosmos is governed from within by the divine reason that pervades it (Hard, 2015, p. xx).

The main part of Aratos' work describes the stars and their successive appearances in the sky. He only mentions the planets to dismiss their study:

> Mixed with them are five other stars [planets], in no way like them, that whirl all through the twelve figures [of the Zodiac]. Not by looking at other stars could you mark the paths of these, since all move about. Long are the periods of their revolutions, and very far apart the signs of their conjunctions. Of them I have no longer confidence: may I be competent to tell of the circles of the fixed stars and the signs in heaven. (Aratos, *Phainomena*, 454-461)

We find a similar view in Xenophon's description of Socrates' attitude towards astronomy:

> He bid them also to become experienced in studying the stars – in this too, however, up to the point of being able to judge the time of night, month, and year, for the sake of traveling, sailing, and guarding, and as for the other things that are done by night or by month or by year to have marks to use in distinguishing the periods of the things mentioned. And he said that these things were easy to learn from night hunters and ship pilots as well as many other whose concern it is to know them. But he turned them strongly away from learning astronomy as far as knowing also the beings that do not keep to the same path of revolution, both the planets and the unstable stars [comets], or exhausting themselves by seeking their distances from the earth, their paths and the causes of these things. For he said that he did not see any benefit in these things either (and yet even in these he was not uninstructed), and he said that they too were sufficient to exhaust the life of a human being and to hinder one from many beneficial things. (Xenophon, *Memorabilia*, IV.7.4-5)

The oldest surviving Greek works on mathematical astronomy were written by Autolycos of Pitane, about 320 BCE (James, 1998, p. 21). They do not analyze the motion of the planets – their subject is the rotation of the sphere of the stars, and the rising and setting of constellations – that is, the mathematical treatment of the same subject covered by Aratos' work.

Indeed, the Greek word ἄστρων (*astron*) was generally applied to the Sun, the Moon and to the constellations, not to an isolated star of to the planets, which were not a concern for the early Hellenic ἀστρονομία (Tannery, 1893, p. 4). Plato, in the *Republic*, also emphasizes the practical side of astronomy when discussing the useful sciences that should be studied:

> [Socrates:] "Shall we set down astronomy as a third [after arithmetic and geometry], or do you dissent?" "I certainly agree," he [Glaucon] said; "for quickness of perception about the seasons and the courses of the months and the years is serviceable, not only to agriculture and navigation, but still more to the military art". (Plato, *Republic* VII, 527d)

Aratos' poem includes technical information – such as the celestial axis and poles –, mythological aspects and practical uses of astronomical knowledge, such as the use of the two Bears to guide the sailors:

> They [the stars], all alike, many though they be and other star in other path, are drawn across the heavens always through all time continually. But the Axis shifts not a whit, but unchanging is forever fixed, and in the midst, it holds the earth in equipoise, and wheels the heaven itself around[5]. On either side the Axis ends in two Poles, but thereof the one is not seen, whereas the other faces us in the north high above the ocean. Encompassing it two Bears [Ursa Major and Minor] wheel together – wherefore they are also called the Wains. Now they ever hold their heads each toward the flank of the other, and are borne along always shoulder-wise, turned alternate on their shoulders. If, indeed, the tale be true, from

---

[5] The idea of a spherical rotating heaven was probably introduced in Greek thought by Pythagoras.

Crete they by the will of mighty Zeus entered up into heaven, for that when in olden days he played as a child in fragrant Dicton, near the hill of Ida, they set him in a cave and nurtured him for the space of a year, what time the Dictaean Curetes were deceiving Cronus. Now the one [of the two Bears] men call by name Cynosura [Ursa Minor] and the other Helice [Ursa Major]. It is by Helice that the Achaeans on the sea divine which way to steer their ships, but in the other the Phoenicians put their trust when they cross the sea. But Helice, appearing large at earliest night, is bright and easy to mark; but the other is small, yet better for sailors: for in a smaller orbit wheel all her stars. By her guidance, then, the men of Sidon steer the straightest course. (Aratos, *Phainomena*, 19-44)

Some of the Greeks ascribed the discovery of the spherical form of the Earth to Pythagoras (sixth century BCE), others to Parmenides (fifth century BCE). According to Sulpicius Gallus, Thales of Miletus was the first to represent the heavens with a sphere (James, 1998, pp. 47, 90). The cosmological view of the Greeks, with a spherical universe with the Earth at its center, entailed a sense of order and aesthetic beauty associated to the very concept of κόσμος (Taub, 2003, p. 13).

It is likely that Aratos' *Phainomena* soon became a standard textbook, and it was supplemented by commentaries and additional works. The understanding of Aratos' work required a knowledge of the constellations, that were probably studied using celestial globes and charts (Hard, 2015, p. xxii).

## 7. ASTRONOMY AND METEOROLOGY

A very important part of Aratos' work is the description of the relations between the weather and the heavenly signs. This was a very important subject to be studied by sailors, for instance:

Those twelve signs of the Zodiac are sufficient to tell the limits of the night. But they to mark the great year – the season to plough and sow the fallow field and the season to plant the tree – are already revealed of Zeus and set on every side. Yea, and on the sea, too, many a sailor has marked the coming of the stormy tempest, remembering either dread Arcturus or other stars that draw from ocean in the morning twilight or at the first fall of night. (Aratos, *Phainomena*, 740-747)

In the Odyssey and the Iliad, the meteorological phenomena are often linked to gods. Zeus is "cloud gatherer", and Athena both sends and stalls winds (Taub, 2003, p. 5). However, the ancient practice of using the risings and settings of stars to mark out the seasons and to predict weather phenomena was based on ideas that the motions of the heavens are related to terrestrial effects, particularly weather (Taub, 2003, p. 8). Some of the weather signs were related to the Moon:

Scan first the horns on either side the Moon. For with varying hue from time to time the evening paints her and of different shape are her horns at different times as the Moon is waxing – one form on the third day and other on the fourth. From them thou canst learn touching the month that is begun[6]. If she is slender and clear about the third day, she heralds calm: if slender and very ruddy, wind; but if thick and with blunted horns she shows but a feeble light on the third and fourth night, her beams are blunted by the South wind or imminent rain. If on the third night neither horn nod forward or lean backward, if vertical they curve their tips on either side, winds from the West will follow that night. But if still with vertical crescent she brings the fourth day too, she gives warning of gathering storm. If her upper horn nod forward, expect thou the North wind, but if it leans backward, the South. But when on the third day a complete halo, blushing red, encircles her, she foretells storm and, the fierier her blush, the fiercer the tempest. (Aratos, *Phainomena*, 778-799)

---

[6] The Greek month began on the New Moon.

A similar account of the relation between the astronomical signs and weather occurs in a work ascribed to Theophrastos (Tannery, 1893, p. 17).

The Greeks preferred not to sail during the stormy winter months. The best conditions for sailing were from late May to mid-September, that is, after the rising of the Pleiades to the rising of Arcturus (Irby, 2016, v. 1, p. 861). The four main winds of the Greek tradition, mentioned by Homer and other ancient authors, were Boreas (from the North), predominant in winter, bringing fine weather, but strong and violent; Notos (from the South), a dangerous, stormy wind associated to late summer and autumn storms; Zephyros (from the West), bringing spring and early summer breezes; and Euros (from the East) (Bunbury, 1879, vol. 1, p. 36; Daremberg, 1885, p. 329). In ancient Greek mythology, they were regarded as wind gods; in the Odyssey, they are horses kept in the stables of the storm god Aeolus. During the year, each of those main winds was announced by specific constellations:

> Beneath the head of Helice are the Twins [Gemini]; beneath her waist is the Crab [Cancer]; beneath her hind feet the Lion [Leo] brightly shines. There is the Sun's hottest summer path. Then the fields are seen bereft of corn-ears, when first the Sun comes together with the Lion. Then the roaring Etesian winds[7] fall swooping on the vast deep, and voyaging is no longer seasonable for oars. Then let broad-beamed ships be my choice, and let steersmen hold the helm into the wind. (Aratos, *Phainomena*, 147-156)

## 8. MEASUREMENT OF TIME

Astronomy also had a fundamental role in the measurement of the parts of the day. Instead of dividing the time between one midnight and the next one into twenty-four equal parts, as we do, the Greeks divided the time between sunrise and sunset in

---

[7] The name "etesian" is related to ἔτος, year. The Etesian winds are the strong North winds.

twelve hours which changed their length through the year; and night was also divided in twelve hours (Evans, 1998, p. 95). The Egyptians were probably the first people to divide the night in 12 parts, as this was recorded since 2150 BCE (Hannah, 2005, p. 87).

Since it depended on sunrise and sunset, the measurement of time was directly related to the motion of the Sun. There were several ways of measuring times smaller than a day – such as water clocks (clepsydrae) and other devices; but since the hour was defined as the 12th part of the bright part of the day (between sunrise and sunset), the most useful and precise clocks of Antiquity used during daytime were sundials. For all practical uses, no other instrument could provide an adequate measure of time, because only sundials provided variable hours, smaller during the winter and larger during the summer.

According to Herodotus, the division of the duration of the day in 12 parts, the device for observing the shadow of the Sun (the *gnomon*) and the first spherical sundial (the *polos*) were imported by the Greeks from Babylon (Tannery, 1887, p. 82). The *gnomon* can be used to find the equinoxes, the obliquity of the ecliptic, the height of the celestial pole (and, therefore, the geographical latitude). Some ancient sources ascribe to Anaximander of Miletus (c. 610-546 BCE) the erection of a *gnomon* in Lacedaemon; and to Aristarchus of Samos the construction of a plane sundial, that requires advanced mathematical analysis of the motion of the Sun (*ibid.*, pp. 82-83). The design of a plane sundial is a very complicated subject. Besides ascertaining the cardinal directions, it requires knowledge of the latitude of the place where it is built, and uses a set of complex geometrical procedures.

During the night, the measurement of time employed the observation of the stars. In each night, six signs of the Zodiac rise and set; hence, the appearance or disappearance of each sign corresponds to two hours. To ascertain the time, it was necessary both to observe the rising zodiacal sign on the eastern horizon; and to know the current sign of the Sun (Evans, 1998, p. 95).

The beginning of the night corresponded to the rising of the sign opposite to the one occupied by the Sun, and so on. From the fifth century BCE on, the information about the position of the Sun would be known to the average person; in some towns, *parapegmata* (public calendars) displayed the current place of the Sun in the Zodiac, along with other information (Evans, 1998, p. 98)

## 9. ASTRAL MYTHOLOGY

The sky was regarded by several ancient civilizations – including the Greek one – as a sacred space. The Pythagoreans held that the heavenly bodies are divine, moved by the ethereal soul which informs the universe and is akin to man's soul; Plato regarded the stars are visible gods (Cumont, 1912, p. 39).

We have remarked that a few stars and constellations had already names and mythological meaning in the early Greek culture, as seen in the works of Homer and Hesiod. Most astral myths were developed, however, after the fifth century BCE, flourishing later in the Hellenistic period. Sophocles and Euripedes are cited as sources for the myths related to the constellations of the Perseus-Andromeda group, and Pherecydes for the mythology of the Hyades and Ariadne's crown, for instance (Hard, 2015, pp. xiv-xv). In a few centuries, the splendid figures set in the heavenly sacred space were identified with people, creatures and things from myth and legend. They had been placed or depicted in the sky, usually by gods, in a process called *katasterismos*.

In Aratos' *Phainomena* we can find the names and a description of symbolic imagens covering the sky, including almost all the constellations of Ptolemy's later catalogue (Bouché-Leclercq, 1899, p. 62). His source was Eudoxos' fourth century work, and therefore at this time there was a definite identification of the observable stars. However, Aratos was not especially interested in myths – he describes only a few stories concerning the constellations, his main interest being their positions, risings and settings. Not much later than Aratos,

Eratosthenes (ca. 276-194 BCE) wrote a lost astronomical work, called *Katasterismoi*, that was probably the most complete early collection of myths related to the stars, besides describing the stars in each constellation (Hard, 2015, pp. xvii-xviii). In many cases, astral lore was intimately linked to the influences associated to each constellation or star. We have remarked that the Dog Star (Sirius) brought strong heat; Eratosthenes presented a myth for the Dog Star that explained this characteristic (Hard, 2015, p. 117).

The content of Eratosthenes' work is known through a Latin manual of astronomy ascribed to Gaius Julius Hyginus (ca. 64 BCE-17 CE), the *Astronomia*, or *Poeticon astronomicon*. It contains the basic (non-mathematical) ancient astronomical knowledge, but was mostly devoted to star myths, often with variants and alternative stories (Hard, 2015, pp. xvii; xxvi-xxvii). This canon of astral mythology became part of the general Hellenistic and Roman cultures. Educated people would be expected to have a broad knowledge of astronomy and of the celestial myths.

## 10. MEDICINE AND THE STARS

Another realm of human concern that was related to the heaven, in the ancient Greek culture, was medicine. In Plato's *Symposium*, Eryximachus (a physician) explained the relevance of equilibrium between the opposite powers for health, and its relation with the influences of the stars that bring the different seasons:

> Thus, in music and medicine and every other affair, whether human or divine, we must be on the watch as far as may be for either sort of love; for both are there. Note how even the system of the yearly seasons is full of these two forces; how the qualities I mentioned just now, heat and cold, drought and moisture, when brought together by the orderly Love, and taking on a temperate harmony as they mingle, become bearers of ripe fertility and health to men and animals and plants, and are guilty of no wrong. But when the wanton-

spirited Love gains the ascendant in the seasons of the year, great destruction and wrong does he wreak. For at these junctures are wont to arise pestilences and many other varieties of disease in beasts and herbs; likewise hoar-frosts, hails, and mildews, which spring from mutual encroachments and disturbances in such love-connections as are studied in relation to the motions of the stars and the yearly seasons by what we term astronomy. So further, all sacrifices and ceremonies controlled by divination, namely, all means of communion between gods and men, are only concerned with either the preservation or the cure of love. (Plato, *Symposium*, 187e-188c)

The relation between health, the seasons and winds brought by the stars is discussed at length in the treatise *On airs, waters, and places* of the Hippocratic corpus.

Whoever would study medicine aright must learn of the following subjects. First, he must consider the effect of each of the seasons of the year and the differences between them. Secondly, he must study the warm and cold winds, both those which are common to every country and those peculiar to a particular locality. [...] With the passage of time and the change of the seasons, he would know what epidemics to expect, both in the summer and in the winter, and what particular disadvantages threatened an individual who changed his mode of life. Being familiar with the progress of the seasons and the dates of rising and setting of the stars, he could foretell the progress of the year. Thus, he would know what changes to expect in the weather and not only would he enjoy good health himself for the most part but he would be very successful in the practice or medicine. (Hippocrates, *On airs, waters, and places*, 1-2)

This treatise is full of specific predictions, such as these:

And respecting the seasons, one may judge whether the year will prove sickly or healthy from the following observations. If the appearances connected with the rising and setting stars be as they should be; if there be rains in autumn; if the winter be mild, neither very tepid nor unseasonably cold, and if in spring the rains be seasonable, and so also in summer, the year is likely to prove healthy. But if the winter be dry and northerly, and the spring showery and southerly, the summer will necessarily be of a febrile character, and give rise to ophthalmies and dysenteries. For when suffocating heat sets in all of a sudden, while the earth is moistened by the vernal showers, and by the south wind, the heat is necessarily doubled from the earth, which is thus soaked by rain and heated by a burning sun, while, at the same time, men's bellies are not in an orderly state, nor the brain properly dried; for it is impossible, after such a spring, but that the body and its flesh must be loaded with humors, so that very acute fevers will attack all, but especially those of a phlegmatic constitution. Dysenteries are also likely to occur to women and those of a very humid temperament. And if at the rising of the Dog star rain and wintery storms supervene, and if the etesian winds blow, there is reason to hope that these diseases will cease, and that the autumn will be healthy; but if not, it is likely to be a fatal season to children and women, but least of all to old men; and that convalescents will pass into quartans, and from quartans into dropsies; but if the winter be southerly, showery and mild, but the spring northerly, dry, and of a wintry character, in the first place women who happen to be with child, and whose accouchement should take place in spring, are apt to miscarry; and such as bring forth, have feeble and sickly children, so that they either die presently or are tender, feeble, and sickly, if they live. Such is the case with the women. The others are subject to dysenteries and dry ophthalmies, and some have catarrhs beginning in the head and descending to the lungs. Men of a phlegmatic temperament are likely to have dysenteries; and women, also, from the humidity of their nature, the phlegm descending downwards from the brain; those who are bilious, too, have dry ophthalmics from the heat and dryness of their flesh; the aged, too, have catarrhs from their flabbiness and melting of the veins, so that some of them die suddenly and some become paralytic on the right side or the left. (Hippocrates, *On airs, waters, and places*, 10)

The Hippocratic treatise *On Regimen* presents a simple division of the year, not unlike our modern one: "I divide the year into four parts, which most people know best: winter, spring, summer, autumn. And winter lasts from the setting of the Pleiades until the spring equinox; spring, from the spring equinox until the rising of the Pleiades; summer from the Pleiades until the rising of Arcturus; and autumn from Arcturus until the setting of the Pleiades" (Hulskamp, 2011, pp. 151-152)[8]. According to Hippocrates, epidemics were brought by particular weather conditions related to the seasons of the year. For instance:

In Thasus, a little before and during the season of Arcturus, there were frequent and great rains, with northerly winds. About the equinox, and till the setting of the Pleiades, there were a few southerly rains: the winter northerly and parched, cold, with great winds and snow. Great storms about the equinox, the spring northerly, dryness, rains few and cold. About the summer solstice, scanty rains, and great cold until near the season of the Dog-star. After the Dog-days, until the season of Arcturus, the summer hot, great droughts, not in intervals, but continued and severe: no rain; the Etesian winds blew; about the season of Arcturus southerly rains until the equinox. In this state of things, during winter, paraplegia set in, and attacked many, and some died speedily; and otherwise the disease prevailed much in an epidemical form, but persons remained free from all other diseases. Early in the spring, ardent fevers commenced and continued through the summer until the equinox. Those then that were attacked immediately after the commencement of the spring and summer, for the most part recovered, and but few of them died. But when the autumn and the rains had set in, they were of a fatal character,

---

[8] The first recorded Greek observation of the summer solstice was made in 432 BCE by Meton and Euctemon (James, 1998, p. 20).

and the greater part then died. (Hippocrates, *Of the Epidemics*, I.2.7-8)

Besides the general effects of the weather, *On airs, waters, and places* also states that some specific days of the year are particularly dangerous:

> Whoever studies and observes these things may be able to foresee most of the effects which will result from the changes of the seasons; and one ought to be particularly guarded during the greatest changes of the seasons, and neither willingly give medicines, nor apply the cautery to the belly, nor make incisions there until ten or more days be past. Now, the greatest and most dangerous are the two solstices, and especially the summer, and also the two equinoxes, but especially the autumnal. One ought also to be guarded about the rising of the stars, especially of the Dog star, then of Arcturus, and then the setting of the Pleiades; for diseases are especially apt to prove critical in those days, and some prove fatal, some pass off, and all others change to another form and another constitution. (Hippocrates, *On airs, waters, and places*, 11)

The crisis of a disease is a decisive moment: either the disease overcomes the patient, of the patient overcomes the disease, or the disease changes into something completely different (Hulskamp, 2011, p. 158). The Hippocratic doctrine of critical days was closely similar to the astrologers' beliefs. It accepted that there were specific days or hours that were of crucial significance in determining the outcome of an illness. This doctrine was later developed in Galen's medical astrology (Nutton, 2013, p. 275).

## 11. MAN, THE MICROCOSM

The deeper relation between the heavens and the human body was expressed as the view, ascribed to Democritus, that the individual is a *mikros kosmos*; and this analogy between man and the cosmos appears in the Hippocratic treatise *On Regimen*. That work establishes relations between the appearance of stars, the seasons, the weather and its effect on the body, and the adjustments of regimen required at each time. It also uses astronomical phenomena in *diagnosis, prognosis* and treatment of diseases (*prodianosis*) (Hulskamp, 2011, pp. 163-164).

The structure of the human body is related to the Earth and the universe, "a copy of the whole, the small after the manner of the great, and the great after the manner of the small" (Hippocrates, *Regimen* I.10, *apud* Hulskamp, 2011, p. 164). The belly is like the sea, the stomach and the lungs correspond to the earth; the peritoneum is associated to the Moon, the stars to the peripheral circuit close to the skin, the Sun to the intermediate circuit – probably the heart – where the hottest and strongest fire is located and the soul moves. Health problems can be diagnosed by dreams related to the several parts of the universe. Problems associated to the stars indicate that the outer bodily circuit has been affected, and it should be purged outward, through the skin; when it concerns the Moon, there is something wrong in the hollow parts of the body, and purgation should be directed inward; when the problem is related to the Sun, purgation in both directions should take place (Hulskamp, 2011, p. 166).

> Whenever a heavenly body appears to fall away from its orbit, should it be pure and bright, and the motion towards the east, it is a sign of health. For whenever a pure substance in the body is secreted from the circuit in the natural motion from west to east, it is right and proper. In fact, secretions into the belly and substances disgorged into the flesh all fall away from the circuit. But whenever a heavenly body seems to be dark and dull, and to move towards the west, or into the sea, or into the earth, or upwards, disease is indicated. When the motion is upwards, it means fluxes of the head; when into the sea, diseases of the bowels; when into the earth, most usually tumors growing in the flesh. In such cases it is beneficial to reduce food by one-third and to take an emetic, to be followed

by a gradual increase of food for five days, the normal diet being resumed in another five. Another emetic should be followed by the same gradual increase. Whenever a heavenly body seems to settle on you, if it be pure and moist, it indicates health, because what descends from the ether on to the person is pure, and the soul too sees it in its true character as it entered the body. But should the heavenly body be dark, impure and not transparent, it indicates disease caused neither by surfeit nor by depletion, but by the entrance of something from without. It is beneficial in this case to take sharp runs on the round track, that there may be as little melting of the body as possible, and that by breathing as rapidly as possible the patient may secrete the foreign body. After these runs let there be sharp walks. Diet to be soft and light for four days. (Hippocrates, *Regimen*, IV.89; Hippocrates, 1959, pp. 433-435)

## 12. GEOGRAPHY AND ASTRONOMY

Some ancient technical developments, such as the study of geography and the manufacture of maps, also required astronomical knowledge.

Homer did not mention the four cardinal points; he vaguely referred to those directions as "towards the dawn and the Sun" (east) and "to darkness" (west); and he also mentioned the four main winds, Boreas, Notos, Zephyros and Euros (corresponding to north, south, west and east), without stating their directions (Bunbury, 1879, vol. 1, pp. 35-36, 77).

By the fourth century BCE, Greek philosophers and astronomers accepted that the Earth was a sphere at rest in the middle of a finite universe, and that the fixed stars were bound to a sphere that rotates uniformly around a fixed axis (James, 1998, pp. 19-20). This seemingly simple geometrical view of the universe allowed the Greek to develop sophisticated theories of the motion of the stars, and also a mathematical geography grounded upon the observation of the sky.

Nowadays, the use of a magnetic compass for ascertaining the geographical directions is so common that we forget that this instrument was unknown in Europe before the 12th century and that before that time the concept of the cardinal directions was an astronomical one. The geographical directions were defined in Antiquity by taking into account the structure of the sky: the north and south poles were the extremities of the axis of the celestial sphere, and the north-south direction could be ascertained either by the observation of the night sky (the circumpolar stars) or during daytime by the study of the motion of the Sun, with the use of a *gnomon*, possibly introduced in Greece by Anaximander (Bunbury, 1879, vol. 1, p. 122). This was known by all educated people, in Antiquity.

One of the earliest arguments for the spherical form of the Earth was that different constellations could be observed from different geographical places. At first, this was a qualitative observation made by travelers, but Theodosius of Bithynia (2nd century BCE) wrote a work *On geographical places*, where he discussed the appearances of the stars as seen from different places, taking into account the spherical shapes of the sky and the Earth (James, 1998, p. 24).

The ascertaining of the geographical position of a place on the spherical Earth depends on the determination of its latitude and longitude. The observed angular height of the North celestial pole, relative to the horizon, varies depending on the geographical position of the observer, due to the spherical shape of the Earth. This observable angle allowed the determination of the latitude of the observer on the Earth. It could also be ascertained by the study of the shadows produced by the Sun – especially during the equinoxes or the solstices (Bunbury, 1879, vol. 1, p. 632). So, latitude was ascertained by astronomical observations. It was much more difficult to find the geographical longitudes. Sometimes it was found by direct evaluation of distances and by computing the angular distance between the corresponding meridians. It was possible, however, to evaluate large longitude differences by the timing of lunar eclipses, as described by Hipparchus (Bunbury, 1879, vol. 1, p. 633).

Nowadays, we think about the meridians and the parallel circles (equator, tropics and polar circles) as imaginary lines upon the surface of the Earth. However, both their early definitions and the method of finding them were astronomical. The meridians were the circles of the heavenly sphere that passed by the celestial poles. The equator was the circle drawn in the sky by the Sun during the equinoctial days. The tropics were the circles described by the Sun during the solstices. According to Plutarch (*De Placitis Philosophorum* 2.12 and 3.14; see Strabo, 1917, vol. 1, p. 361), Thales and Pythagoras defined those circles and divided the heaven in five zones (Bunbury, 1879, vol. 1, p. 120). The corresponding circles and divisions of the Earth came afterwards, and were defined relative to the celestial ones. Poseidonius ascribed the later division to Parmenides (Strabo, *Geography*, II.2; see Strabo, 1917, vol. 1, p. 361; Bunbury, 1879, vol. 1, p. 125). The surface of the Earth was later divided in seven narrower parallel bands, the "climes" (κλίματα), perhaps by Eratosthenes, or Hipparchus (Shcheglov, 2006).

The lengths of daytime and night vary, depending on the time of the year and the latitude of the observation place[9]. The largest difference occurs during the solstices. The person who first established the ratio of the durations of day and night and latitude is unknown; but it was used by Hypsicles in Alexandria, in the second century BCE (Kelley & Milone, 2011, p. 89). Theodosius of Bithynia (2nd century BCE), in his work *On days and nights*, developed advanced computations on the duration of days and nights, and their relation to the geographical latitude (Tannery, 1893, p. 42).

The phenomenon of variation of the length of days and nights was strikingly observed by the famous traveler Pytheas of Massalia (fourth century BCE) who sailed from the

---

[9] According to the older definition of the hour, every day and every night had exactly 12 hours. The astronomers introduced an artificial hour, by dividing the time between two successive midnights into 24 parts. Using this new definition of hour, the durations of day and night may be different.

Mediterranean to Iberia and the Keltic territory, and then to British islands, measuring the latitudes of the places he reached, up to the northern tip of Scotland. He also reached land to the north of Scotland, perhaps on the Norwegian coast (Roller, 2006, p. 69-75). He may be the first Mediterranean to see and to describe the summer night of high latitudes, where the Sun shines dimly all night, moving from west to east.

Pytheas described that the tides of the ocean increased as the Moon became full, and decreased as it waned (Bunbury, 1879, vol. 1, p. 600). This is not a correct description, but it is one of the earliest accounts of the relation between the tides and the Moon, that were later described by Aetius, Stobaeus, Posidonius, Pliny the Elder and Ptolemy; it was incorporated in the popular lore, establishing one additional important influence of the celestial bodies.

## 13. ASTROLOGY

Since the time of Hesiod, the Greeks accepted several types of celestial influence upon the Earth. This is the main concept supporting astrology. However, the idea of predicting the life of an individual by the study of the celestial configuration at the moment of his birth (genethlialogy) did not exist among the Greeks, at that time. This approach to astrology appeared in Babylon around the fifth century BCE – the earliest known horoscope being from 410 BCE (Beck, 2007, pp. 13-14). Babylonian horoscopes described the position of the ascendant sign of the Zodiac – that is, the one that is rising at east, as viewed from the place where the person is born –, the positions of the Sun, the Moon and, sometimes, of the planets. The earliest Babylonian horoscopes only contained a description of the sky, without interpretation (Barton, 1994, p. 14). Others provided an explanation of the astrological influences, such as this example, from 235 BCE:

> Year 77 [of the Seleucid Era, month] Siman, [from?] the
> 4th [day until? some? time?] in the last part of the night off?]

the fifth [day], Aristocrates was born. That day, Moon in Leo. Sun in 12;30° in Gemini. The Moon set its face from the middle towards the top; [the relevant omen reads:] 'If, from the middle towards the top, it [i.e., the Moon] sets its face, [there will ensue] destruction.' Jupiter [...] in 18° Sagittarius. The place of Jupiter [means]: [his life? will be] regular, well; he will become rich, he will grow old, [his] days will be numerous [literally, long]. Venus in 4° Taurus. The place of Venus [means]: Wherever he may go, it will be favorable [for him]; he will have sons and daughters. Mercury in Gemini with the Sun. The place of Mercury [means]: the brave one will be first in rank, he will be more important than his brothers, [...] Saturn; 6° Cancer. Mars: 24° Cancer [...] the 22nd and 23rd of each month [...] (Barton, 1994, p. 16-17)

In Greece, up to Socrates' time, the planets had been neglected because of their irregular motion; they were neither described nor received individual names (Barton, 1994, p. 21). Plato, however, was already paying attention to them, and in the *Timaeus* he regarded them as divine beings – as was the case in Babylon (Barton, 1994, p. 22). The main character of this Platonic dialogue, called *Timaeus*, was a Pythagorean; and the followers of Pythagoras held that the heavenly bodies are divine, moved by the ethereal soul which informs the universe and is akin to man's soul (Cumont, 1912, p. 39). Some authors find in Plato's *Timaeus* a philosophical inspiration for the development of Greek astrology, since it describes all the fixed and moving stars, and even the Earth, as "living and immortal gods", endowed with body and soul; and that the Demiurgue ordered them to continue the work of fashioning the world and the mortal beings – including men and their souls (Bouché-Leclercq, 1899, pp. 21-22). Contemplation of the heaven produced not only aesthetic emotions, it was the motivation of the powerful belief in a good, rational and powerful deities that produced order (Cumont, 1909, pp. 264-265).

The earliest Greek names of the planets were merely descriptive of their visible characteristics (e.g., Mars was the "fiery star", because of its red color). After the fourth century BCE the five known planets were named after Greek gods: they became the stars of Hermes, Aphrodite, Ares, Zeus, Kronos, probably because of Babylonian influence (Cumont, 1912, pp. 45-46). Notice that the Sun and the Moon were associated to Helios and Selene, which were secondary deities in the Greek pantheon.

Plato's pupil Eudoxus of Cnidus (408-355 BCE), after travelling to Egypt and Anatolia, acquiring new mathematical and astronomical skills, attempted to explain and predict their motions, using homocentric spheres. According to Seneca, Eudoxus was the first to bring the table of planetary motions from Egypt to Greece (Bouché-Leclercq, 1899, p. 63). Aristotle (384-322 BCE) was strongly influenced by this approach. The Aristotelian approach to the heavens was very different from Plato's; however, his philosophy also had a strong influence in later astrology, because of his theory of the four elements and the four main qualities (dry/moist, hot/cold) that was used by Ptolemy and other authors to explain the astral influences (Bouché-Leclercq, 1899, p. 25).

A third important philosophical influence on Greek astrology was the Stoicism of Zeno of Citium (ca. 334-262 BCE). According to the Stoics, man is a microcosm, the image and abridgement of the world. The stars were regarded as living gods, higher than man. The astral gods had the power and the will of interfering on terrestrial affairs. They acted by universal sympathy on their destiny, also sending signs that can be understood by human beings. This led the Stoics to accept all types of divination, including astrology, because of their religious and philosophical principles (Bouché-Leclercq, 1899, pp. 28-31). After Alexander, the combination of Babylonian religious influences with Stoic philosophy produced a very influent astral religion, which flowered with Posidonius of Apameia (ca. 135-51 BCE) and maintained a strong impact during the Hellenistic and Roman periods (Cumont, 1909, pp. 256-257).

Once this basic philosophy of astrology was accepted, the symbolic meanings of each sign of the Zodiac acquired a new function, acting upon the human beings under their influence. The association between the planets and Greek gods established another important way to interpret the astral influences related to the personal characteristics (including physical, psychological and gender peculiarities) of each deity (Bouché-Leclercq, 1899, pp. 67, 89-101, 132).

The study of the motion of the planets was highly complex, and was studied only by the mathematicians. However, some broad acquaintance with the planets and their main phenomena and influences soon became part of the Greek culture. The analysis of the motion of the stars and planets for drawing horoscopes required the development of an approach that was much more detailed and precise than the one known to former astronomers. Hypsicles (ca. 190-120 BCE) was probably the first Greek author who introduced the Babylonian division of the circle in 360 degrees (and also the day in 360 parts) to develop a more accurate analysis of the celestial motions (Bulmer-Thomas, 1972). Genethlialogy probably migrated from Babylon to the Greek world together with mathematical astronomy[10], and it was fully established in the Hellenistic culture in the second century BCE, at the time of Hipparchus (fl. 150-125 BCE) (Beck, 2007, pp. 15-16). The oldest surviving Greek horoscope was produced on 72 BCE (Barton, 1994, p. 16).

Several ancient astrological manuals were produced from the first century CE onwards, both in Hellenistic Greece and in the Roman Empire (Beck, 2007, p. 39). Some of the best known were written by Marcus Manilius (first century), Ptolemy

---

[10] However, according to Cicero, Eudoxus (possibly the first Greek mathematical astronomer) rejected prophecies based on the day of birth, that is, horoscopes (Barton, 1994, p. 22). The names of two Chaldean astrologers who brought the Babylonian knowledge to the Greek world are known: Berosus (ca. 280 BCE) and Soudines (ca. 238 BCE) (Cumont, 1912, pp. 56-57).

(second century) and Julius Firmicus Maternus (fourth century CE).

## 14. THEORY OF MATTER, METALLURGY AND ALCHEMY

The study of stars and planets also became linked to the practical and theoretical study of matter, in Antiquity. This occurred along several different and independent lines.

Aristotle had combined the four elements of Empedocles (fire, air, water, earth) with Hippocrates' doctrine of the four main influences (dry/wet, hot/cold), providing a fundamental connection between them: fire was dry and hot; air, wet and hot; water, wet and cold; and earth, dry and cold. Ptolemy, in his astrological treatise *Tetrabiblos*, associated Aristotle's ideas to several astronomical phenomena (Beck, 2007, p. 59). Each season, produced by the passage of the Sun by a specific set of zodiacal signs, brings a predominance of one of the four main influences. Spring is wet (becoming hot), summer is hot (becoming dry), autumn is dry (becoming cold), winter is cold (becoming wet). Therefore, those seasons (and the corresponding signs of the zodiac) were respectively associated to the elements air, fire, earth and water.

Another correlation was established by Marcus Manilius, who divided the twelve signs in four groups according to the trine aspect, establishing the following associations (Beck, 2007, p. 60):

- The triangle of fire (hot, dry, male): Aries, Leo, Sagittarius
- The triangle of earth (cold, dry, female): Taurus, Virgo, Capricorn
- The triangle of air (hot, wet, male): Gemini, Libra, Aquarius
- The triangle of water (cold, wet, female): Cancer, Scorpius, Pisces

Ptolemy did not associate the elements to the planets, but described their associations with the four qualities. The active power of the Sun is to heat and, to a certain degree, to dry; that of the Moon, to humidify, but it has a moderate heating power;

Saturn's quality is to cool and, moderately, to dry; that of Mars, to dry and to burn; Jupiter has a temperate active force: it heats and humidifies; Venus is similar to Jupiter, but she warms moderately and chiefly humidifies; Mercury sometimes dries and sometimes humidifies, and changes quickly from one to the other (Ptolemy, *Tetrabiblos* I.4; Ptolemy, 1940, pp. 35-39).

The four basic qualities and the four elements were also related by Ptolemy to other quaternities: the four main winds, the four cardinal directions, the four ages of mankind that had been formerly identified by the Pythagoreans: childhood, like spring; youth, like summer; adulthood, like autumn; and old age, like winter (Beck, 2007, p. 59). Galen (ca. 129-200 CE), who lived in the same period as Ptolemy, improved the Hippocratic analysis of the qualities and bodily humors, creating a new medical theory that included a typology of human physical constitutions and temperaments: sanguine (with predominance of blood), melancholic (black bile), choleric (yellow bile) and phlegmatic (phlegm). This medical theory was not born from astrology, but it was soon linked to celestial influences:

> For in general Saturn causes his subjects to have cold bellies, increases the phlegm, makes them rheumatic, meagre, weak, jaundiced, and prone to dysentery, coughing, raising, colic, and elephantiasis; the females he makes also subject to diseases of the womb. Mars causes men to spit blood, makes them melancholy, weakens their lungs, and causes the itch or scurvy; and furthermore, he causes them to be constantly irritated by cutting or cautery of the secret parts because of fistulas, hemorrhoids, or tumors, or also burning ulcers, or eating sores; he is apt to afflict women furthermore with miscarriages, embryotomies, or corrosive diseases. (Ptolemy, *Tetrabiblos*, 3.12; Ptolemy, 1940, pp. 327-329)

A completely different tradition, with strong Babylonian influence, established relations between the planets, their gods, and seven metals. Both in Mesopotamia and in Egypt, gold was related to the Sun god and was supposed to have life-giving properties; and silver was the metal of the Moon god Sin (Forbes, 1950, pp. 86, 177). Those associations were also accepted in ancient Greece, together with a few others that were not so generally accepted. Aphrodite (and her planet, Venus) was associated to copper, a metal named after the island of Cyprus; and Aphrodite was sometimes called Cypris, because she was born in that island. Besides that, most mirrors were made of bronze, and copper is one of its components. Aphrodite, the beautiful goddess, was associated to mirrors (Cyrino, 2012, pp. 27, 66).

From the second century CE onward, we find an invariable connection in Hellenistic and Roman literature between Sun and gold, Moon and silver, Saturn and lead, Venus and copper, Mars and iron; the later relation between Jupiter and tin, and Mercury and quicksilver, had not been recognized at this time (Lippmann, 1919, p. 217). Those specific identifications between planets, gods and metals were also recognized by the Assyrians (Karpenko, 2003, p. 216).

The lapidary attributed to Damigeron and Evax (perhaps second century BCE), describes stones related to the seven planets and their engraving to produce amulets (Quack, 2001, p. 338). Sometimes, a relation was also established between the planets and precious stones that were placed on a board, to represent them, when casting a horoscope (Evans, 2004):

> A voice comes to you speaking. Let the stars be set upon the board in accordance with [their] nature except for the Sun and the Moon. And let the Sun be golden, the Moon silver, Kronos of obsidian, Ares of reddish onyx, Aphrodite lapis lazuli veined with gold, Hermes turquoise: let Zeus be of stone crystalline; and the horoscope, in accordance with [nature]. (Betz, 1992, p. 312, *apud* Campion, 2012, pp. 157-158)

In the first centuries CE there were many works pointing out relationships between the planets and specific stones, but the correspondences varied widely (Hall, 2007, pp. 41-54). The relations between the planets, their gods and metals had a strong influence in later alchemy.

## 15. CONCLUSION: A SYMBOLIC NETWORK

The introduction of the astrological symbolism – especially that related to the planets – gave rise to a complex network of correspondences between different realms. Instead of the dry Aristotelian quaternities, we find in several Hellenistic authors the description of emblematic homologies such as this:

> First, the all-seeing sun, burning fire and intellectual light, instrument of spiritual perception; in a nativity, it means kingship, authority, intelligence, wisdom, beauty, social respect, revelation of the gods, judgement, reputation, action, authority over the people; it means the father, the master, friendship, noble persons, the honor having his portrait made, his statue done, of using a crown, command upon one's country [and over] the places. The parts of the body that it rules: the head, the sense organs, the right eye, the trunk, the heart, the breath, all perceptions, the nerves. Of the goods, it rules gold; of fruits, wheat and barley. It is of the day sect. Its color, like wine; its taste, bitter. (Valens, *Anthologies* I.1; Valens, 1989, p. 27)

In Marcus Manilius' *Astronomica* we find the classical distribution of the parts of the human body to different signs of the Zodiac:

> And just as the human frame is apportioned among the signs, a and the protection they afford, though collectively extending over the whole body, is in addition exercised separately over the limbs allocated among them. The Ram is attached to the head, the Bull to the neck; the arms are reckoned as under the Twins' domain, the breast under the Crab's; the shoulders appeal to you, Nemean [Leo], and to you, Maiden [Virgo], the belly; the Balance attends the loins, and the Scorpion is lord of the groin; the Archer [Sagittarius]

has bestowed his love upon the thighs, Capricorn upon the knees, whilst the Youth [Aquarius] is protector of the shanks and the Fishes of the feet. So in like manner do different signs lay claim to different lands. (Manilius, *Astronomica* IV.701-710; Manilius, 1977, p. 279)

Those correlations between the human body and the macrocosm, accepted by all later authors, became of utmost importance in astrological medicine (*iatromathematics*), up to Modernity (Beck, 2007, p. 69).

According to Ptolemy, medical astrology was first developed in Egypt, and it seems to have been a fairly sophisticated discipline. The physician-astrologer would examine the patient and also cast his nativity, which would give him additional information about the patient's state of health. The stars might tell him about the weak points in the patient's organism, or they might warn him of an impending crisis. If, after having made a prognosis, the physician-astrologer hesitated to choose between two types of treatment, the stars might indicate which one was preferable. (Luck, 2006, p. 416)

Another relevant development was the establishment of relations between the sky and the regions of the Earth. The earliest known Hellenistic author who developed an astrological approach to geography was Serapion (second century BCE), probably a pupil of Hipparchus. He established a relation between the seven geographical climes and the planets that determined the character of nature and man in each of them. His account was reproduced by Nigidius Figulus and influenced Pliny the Elder (Honigmann, 1929, p. 30; cf. Cumont, 1930, p. 231; Sarton, 1930, 272)

Both Ptolemy and Manilius also described the relations between the regions of the Earth and the signs of the Zodiac.

God has divided the world into portions, distributing it among the individual signs. To each guardian power he has given a special region of the world to rule, bestowing also the peoples and mighty cities proper to them, wherein the signs should claim their predominant influences. (Manilius, *Astronomica*, IV.697-700; Manilius, 1977, p. 279)

The ancient astrologers attempted to describe and to explain the character of the people inhabiting each country, by the interpretation of the influences of the planets and signs upon each region – a combination of astral geography and ethnography:

Of these same countries Britain, (Transalpine) Gaul, Germany, and Bastarnia are in closer familiarity with Aries and Mars. Therefore, for the most part their inhabitants are fiercer, more headstrong, and bestial. But Italy, Apulia, (Cisalpine) Gaul, and Sicily have their familiarity with Leo and the Sun; wherefore these peoples are more masterful, benevolent, and co-operative. Tyrrhenia, Celtica, and Spain are subject to Sagittarius and Jupiter, whence their independence, simplicity, and love of cleanliness. The parts of this quarter which are situated about the center of the inhabited world, Thrace, Macedonia, Illyria, Hellas, Achaia, Crete, and likewise the Cyclades, and the coastal regions of Asia Minor and Cyprus, which are in the south-east portion of the whole quarter, have in addition familiarity with the south-east triangle, Taurus, Virgo, and Capricorn, and its co-rulers Venus, Saturn, and Mercury. As a result, the inhabitants of those countries are brought into conformity with these planets and both in body and soul are of a more mingled constitution. They too have qualities of leadership and are noble and independent, because of Mars; they are liberty-loving and self-governing, democratic and framers of law, through Jupiter; lovers of music and of learning, fond of contests and clean livers, through Venus; social, friendly to strangers, justice-loving, fond of letters, and very effective in eloquence, through Mercury; and they are particularly addicted to the performance of mysteries, because of Venus's occidental aspect. (Ptolemy, *Tetrabiblos* II.3; Ptolemy, 1940, pp. 135-137)

The seven planets were incorporated into time reckoning by the introduction of the week and the weekdays. This was a late innovation in the Roman calendar, introduced in the time of Augustus (first century CE). Before that, the Romans used an eight-day "week", based on the periodicity of *nundinae*, the market days (Hannah, 2005, pp. 102, 141). The names of the weekdays are derived originally from those of the seven planet-gods (including Sun and Moon), beginning with Saturday: Saturn, Sun, Moon, Mars, Mercury, Jupiter and Venus. Each hour of each weekday was associated to a planet, following the Chaldean order of the planets, from the most distant ones to the closest ones (Saturn, Jupiter, Mars, Sun, Venus, Mercury and Moon). The first hour of Saturday was ruled by Saturn; and the following ones by the successive planets (Hannah, 2005, pp. 141-142). According to Cassius Dio (ca. 155–235 CE), the association of the planets with the hours of the day came before and explained their relation with the weekdays:

> The custom, however, of referring the days to the seven stars called planets was instituted by the Egyptians, but is now found among all mankind, though its adoption has been comparatively recent; at any rate the ancient Greeks never understood it, so far as I am aware. But since it is now quite the fashion with mankind generally and even with the Romans themselves, I wish to write briefly of it, telling how and in what way it has been so arranged. [...] If you begin at the first hour to count the hours of the day and of the night, assigning the first to Saturn, the next to Jupiter, the third to Mars, the fourth to the Sun, the fifth to Venus, the sixth to Mercury, and the seventh to the Moon, according to the order of the cycles which the Egyptians observe, and if you repeat the process, you will find that the first hour of the following day comes to the Sun. And if you carry on the operation throughout the next twenty-four hours in the same manner as with the others, you will dedicate the first hour of the third day to the Moon, and if you proceed similarly through the rest, each day will receive its appropriate god. This, then, is the

tradition. (Dio, *Roman history*, 37.18-19; Dio, 1914, vol. 3, pp. 131-133)

The preparation of talismans and other types of magic rituals required the knowledge of the ruling planet-god of each hour of each day.

Oriental influences helped to produce astral religions in the Hellenistic and Roman worlds. Mythra was originally a Persian god, but it also became popular in Greece (Cumont, 1903, pp. 8-9). It was associated to Helios, but his Persian name was never replaced in the liturgy by a translation, as had been the case with the other divinities worshipped in the Mysteries (*ibid.*, p. 20). The central iconography of Mythraism was the Tauroctony, in which the divine hero Mythra was shown killing a bull (Taurus), surrounded by other constellational images, including Canis, Hydra, Crater and Corvus (Campion, 2012, p. 158). The obvious astronomical correlations were sometimes expanded by introducing the symbols of the Zodiac, and two youths holding the upright and the inverted torch, that represented the northern and southern paths of the Sun during each half year. The artistic creation of Mythra Tauroctonos was probably a Hellenistic production by a sculptor from Pergamon, in the second century BCE (Cumont, 1903, pp. 24, 124).

The Mythraic mysteries were concerned with the individual salvation, regarded as a return to the celestial world. A similar idea was already extant in Plato's *Timaeus* that described that the individual souls originated in the stars and descended to the Earth, leaving divinity behind, but returning to it after death (Campion, 2012, pp. 153-154). In the mysteries of Mythra, the spiritual ascent to the heavenly realm was represented by an alchemical ladder, described by Origen (ca. 185-254 CE):

> There is a ladder with lofty gates [...] The first gate consists of lead, the second of tin, the third of copper, the fourth of iron, the fifth of a mixture of metals, the sixth of silver, and the seventh of gold. The first gate they assign to Saturn, indicating by lead the slowness of this star; the second to

> Venus, comparing her to the splendor and softness of tin; the third to Jupiter, being firm and solid; the fourth to Mercury, for both Mercury and iron are fit to endure all things, and are money-making and laborious; the fifth to Mars, because, being composed of a mixture of metals, it is varied and unequal; the sixth, of silver, to the Moon; the seventh, of gold, to the Sun, thus imitating the different colors of the two latter. (Origen, *apud* Karpenko, 2003, pp. 216-217)

Notice that the order of the planets, here, is the inverse order of the weekdays. Sunday, over which the Sun presided, was especially holy in Mythra's mysteries (Cumont, 1903, p. 167).

The ladder represented seven steps of initiation in a ritual of spiritual ennoblement; as the soul passes each level (or astronomical sphere) it discards the vices associated with that planet (Campion, 2012, p. 158).

> As the soul traversed these different zones, it rid itself, as one would of garments, of the passions and faculties that it had received in its descent to the earth. It abandoned to the Moon its vital and nutritive energy, to Mercury its desires, to Venus its wicked appetites, to the Sun its intellectual capacities, to Mars its love of war, to Jupiter its ambitious dreams, to Saturn its inclinations. It was naked, stripped of every vice and every sensibility, when it penetrated the eighth heaven to enjoy there, as an essence supreme, and in the eternal light that bathed the gods, beatitude without end. (Cumont, 1903, p. 145)

There is a parallel between this ladder with alchemy, since it presents a series of seven metals, from the lowest (lead) to the highest and most perfect (gold) that became a model for the successive transformations in alchemical theory. This is one of the points where the worlds of esoteric and exoteric alchemy met, and this parallel persisted over the entire period of alchemy. (Karpenko, 2003, p. 217).

Any group of seven things could (and ultimately was) related to the seven planets and their gods, establishing symbolic relations. So, the seven vowels of the Greek alphabet came to correspond to the seven planetary gods (Luck, 2006, p. 51).

An astrological pharmacology and botany was also developed, especially in the so-called Hermetic tradition. Thessalus (first century CE) wrote a treatise describing the powers of herbs, where he stated that he was transmitting a revelation by Asclepius, that had been taught by Hermes Trismegistus. The treatise named seven herbs associated with the planets and twelve plants related to the signs of the Zodiac (Scarborough, 1991, p. 155). One example:

> First named among the plants of the Sun is the "heliotrope" [attracted to the Sun]; yet there are many kinds of "heliotropes", and of all these most efficacious is the one called chicory. Its juice mixed with oil of roses is an ointment. It is suitable for relieving heartburns, and it releases tertians, quartans, and intermittent fevers, and mixed with an equal part of the oil of unripe olives, it stops headaches. If someone looking toward the sunrise smears on the juice of the chicory, invoking the presence of the [god] Helios, and begs to give him praise, he will be most favored among all men on that day. (Scarborough, 1991, p. 155)

Other examples could be added to illustrate the web of relationships that connected in Antiquity the planets, stars and gods to all features of the physical world, including people, plants, animals, stones, etc. (Campion, 2012, p. 160).

All those instances show that in classical antiquity, astronomy had a strong connection with everyday life and was deeply embedded in the philosophical and religious culture. Although this relationship changed during the Middle Ages, due to both Islamic and Christian influences, astronomy remained of paramount importance in the culture and practical life until the changes brought by the Renaissance.

## ACKNOWLEDGMENTS

The author is grateful to the São Paulo State Foundation for the Support of Research (FAPESP) and to the Brazilian Council for Scientific and Technological Development (CNPq) for the for the support received during the elaboration of this research.

## BIBLIOGRAPHICAL REFERENCES

AESCHYLUS. *Aeschylus. Vol. 1. Prometheus Bound.* Translation by Herbert Weir Smyth. Cambridge, MA: Harvard University Press, 1926.

ARATOS. *Eratosthenes and Hyginus. Constellation myths; with Aratos's Phaenomena.* A new translation by Robin Hard. Oxford: Oxford University Press, 2015.

ARATOS. *The Phenomena and Diosemeia of Aratos.* Translated by John Lamb. London: John W. Parker, 1848.

BARTON, Tamsyn. *Ancient astrology.* London: Routledge, 1994.

BECK, Roger. *A brief history of ancient astrology.* Malden: Blackwell, 2007.

BERRY, William Turner; POOLE, Herbert Edmund. *Annals of printing: a chronological encyclopaedia from the earliest times to 1950.* London: Blandford Press, 1966.

BETZ, Hans Dieter. *The Greek magical papyri in translation, including the Demotic spells.* Chicago: University of Chicago Press, 1992.

BOUCHÉ-LECLERCQ, Auguste. *L'astrologie grecque.* Paris: Ernest Leroux, 1899.

BOUTSIKAS, Efronsyni. *Astronomy and ancient Greek cult. An application of archaeoastronomy to Greek religious architecture, cosmologies and landscapes.* PhD dissertation, School of Archaeology and Ancient History, University of Leicester, 2007.

BOUTSIKAS, Efronsyni. Placing Greek temples: an archaeoastronomical study of the orientation of ancient

Greek religious structures. *Archaeoastronomy*, 21: 4-19, 2007.

BROWNE, C. A. The poem of the philosopher Theophrastos upon the sacred art: a metrical translation with comments upon the history of alchemy. *The Scientific Monthly*, 11 (3): 193-214, 1920.

BULMER-THOMAS, Ivor. Hypsicles. Vol. 6, pp. 616-617, in: GILLISPIE, Charles Coulston (ed.). *Dictionary of scientific biography*. New York: Scribner, 1972.

BUNBURY, Edward Herbert. *A history of ancient geography among the Greeks and Romans: from the earliest ages till the fall of the Roman Empire*. London: John Muray, 1879. 2 vols.

CAMPION, Nicholas. *Astrology and cosmology in the world's religions*. New York: New York University Press, 2012.

CLINTON, Kevin. The mysteries of Demeter and Kore. Pp. 342-356, in: OGDEN, Daniel (ed.). *A companion to Greek religion*. Malden: Blackwell, 2007.

CUMONT, Franz. *The mysteries of Mithra*. Translated from the second revised French edition by Thomas J. McCormack. Chicago: Open Court, 1903.

CUMONT, Franz. Le mysticisme astral dans l'Antiquité. *Bulletin de l'Académie Royale de Belgique*, Classe des Lettres et des Sciences Morales et Politiques, 5: 256-286, 1909.

CUMONT, Franz. *Astrology and religion among the Greeks and Romans*. Translated by J. B. Baker. New York: G. P. Putnam's Sons, 1912.

CUMONT, Franz. [Review] Honigmann (Ernst). Die sieben Klimata und die πόλεις επίσημο. *Revue Belge de Philologie et d'Histoire*, 9 (1): 231, 1930.

CYRINO, Monica S. *Aphrodite*. London: Routledge, 2012.

DAVIDSON, James. Time and Greek religion. Pp. 204-218, in: OGDEN, Daniel (ed.). *A companion to Greek religion*. Malden: Blackwell, 2007.

DIO, Cassius. *Dio's Roman history*. Translation by Earnest Cary. Cambridge, MA: Harvard University Press, 1914-1927. 9 vols.

EVANS, James. *The history and practice of ancient astronomy*. New York: Oxford University Press, 1998.

EVANS, James. The astrologer's apparatus: a picture of professional practice in Greco-Roman Egypt. *Journal for the History of Astronomy*, 35 (118): 1-44, 2004.

EVELYN-WHITE, Hugh G. *Hesiod, the Homeric hymns, and Homerica*. London: William Heinemann, 1914.

FORBES, Robert James. *Metallurgy in Antiquity: a notebook for archaeologists and technologists*. Leiden: Brill, 1950.

GAWLINSKI, Laura. Greek calendars. Vol. 1, pp. 891-905, in: IRBY, Georgia L. (ed.). *A companion to science, technology, and medicine in ancient Greece and Rome*. Chichester: John Wiley & Sons, 2016. 2 vols.

HALL, Judy. *The stone horoscope: evidence for continuity of ancient esoteric tradition and practice*. MA in Cultural Astronomy and Astrology. Bath Spa University, 2007.

HANNAH, Robert. *Greek and Roman calendars: constructions of time in the classical world*. London: Duckworth, 2005.

HARD, Robin. *Constellation myths: with Aratus's Phaenomena*. Oxford: Oxford University Press, 2015.

HEATH, Thomas. *Aristarchus of Samos, the ancient Copernicus. A history of Greek astronomy to Aristarchus*. Oxford: Clarendon Press, 1913.

HOMER. *The Odyssey of Homer done into English prose*. Translated by S. H. Butcher and A. Lang. New York: P. F. Collier & Son, 1914.

HOMER. *The Odyssey*. Translated by Edward McCrorie. Introduction and notes by Richard P. Martin. Baltimore: Johns Hopkins University Press, 2004.

HONIGMANN, Ernst. *Die sieben Klimata und die* πόλεις ἐπίσημοι [poleis episemoi]: *eine Untersuchung zur Geschichte der Geographie und Astrologie im Altertum und Mittelalter*. Heidelberg: Carl Winters, 1929.

HULSKAMP, Maithe A. A. Space and body: uses of Astronomy in Hippocratic medicine. Pp. 149-168, in: BAKER, Patricia A.; NIJDAM, Han; LAND, Karine van't (eds.). *Medicine and space: body, surroundings and borders in Antiquity and the Middle Ages*. Leiden: Brill, 2011.

IRBY, Georgia L. Navigation and the art of sailing. Vol. 1, pp. 854-869, in: IRBY, Georgia L. (ed.). *A companion to science, technology, and medicine in ancient Greece and Rome*. Chichester: John Wiley & Sons, 2016. 2 vols.

KARPENKO, Vladimír. System of metals in alchemy. *Ambix*, 50 (2): 208-230, 2003.

KELLEY, David H.; MILONE, Eugene F. *Exploring ancient skies. A survey of ancient and cultural astronomy*. New York: Springer, 2011.

LIPPMANN, Edmund Oskar von. *Entstehung und Ausbreitung der Alchemie; mit einem Anhange: zur älteren Geschichte der Metalle; ein Beitrag zur Kulturgeschichte*. Berlin: Julius Springer, 1919.

LUCK, Georg. *Arcana mundi: magic and the occult in the Greek and Roman worlds*. 2nd ed. Baltimore: Johns Hopkins University Press, 2006.

MANILIUS, Marcus. *Astronomica*. Ed. and trans. George P. Goold. Cambridge, MA: Harvard University Press, 1977.

NUTTON, Vivian. *Ancient Medicine*. 2nd ed. London: Routledge, 2013

ODOM, Robert L. Vettius Valens and the planetary week. *Andrews University Seminary Studies*, 3 (1): 110-137, 1965.

PLATO. *Plato in twelve volumes*. Vol. 9. Translated by Harold N. Fowler. Cambridge, MA, Harvard University Press; London, William Heinemann Ltd. 1925.

PTOLEMY, Claudius. *Tetrabiblos*. Translation by Frank Egleston Robbins. Cambridge, MA: Harvard University Press, 1940.

QUACK, Joachim Friedrich. Zum ersten astrologischen Lapidar im Steinbuch des Damigeron und Evax. *Philologus. Zeitschrift für antike Literatur und ihre Rezeption*, 145 (2): 337-344, 2001.

ROLLER, Duane W. *Through the Pillars of Herakles: Greco-Roman exploration of the Atlantic*. New York: Routledge, 2006.

SARTON, George. [Review] Die sieben Klimata und die πόλεις ἐπίσημοι by Ernst Honigmann. *Isis*, 14 (1): 270-276, 1930.

SCARBOROUGH, John. The pharmacology of sacred plants, herbs, and roots. Pp. 138-174, in: FARAONE, Christopher A.; OBBINK, Dirk (eds.). *Magika Hiera: ancient Greek magic and religion*. New York: Oxford University Press, 1991.

SCULLION, Scott. Festivals. Pp. 190-203, in: OGDEN, Daniel (ed.). *A companion to Greek religion*. Malden: Blackwell, 2007.

SHCHEGLOV, Dmitry. Eratosthenes' parallel of Rhodes and the history of the system of climata. *Klio*, 88 (2): 351-359, 2006.

SHEPPARD, Harry J. Gnosticism and alchemy. *Ambix*, 4: 86-101, 1957.

SHEPPARD, Harry J. The redemption theme and Hellenistic alchemy. *Ambix*, 7 (1): 42-46, 1959.

SHEPPARD, Harry J. Alchemy: origin or origins? *Ambix*, 17 (2): 69-84, 1970.

STRABO. *The geography of Strabo*. Translation by Horace Leonard Jones. Cambridge, MA: Harvard University Press, 1917-1932. 8 vols.

TANNERY, Paul. *Pour l'histoire de la science Hellène: de Thalès a Empédocle*. Paris: Félix Alcan, 1887.

TANNERY, Paul. *Recherches sur l'histoire de l'astronomie ancienne*. Paris: Gauthier-Villars, 1893.

TAUB, Liba. Meteorology. Vol. 1, pp. 232-246, in: IRBY, Georgia L. (ed.). *A companion to science, technology, and*

*medicine in ancient Greece and Rome*. Chichester: John Wiley & Sons, 2016. 2 vols.
VALENS, Vetius. *Anthologies, livre I*. Edition, translation and comments by Joëlle-Frédérique Bara. Leiden: Brill, 1989.

# THE TRANSFORMATION OF ASTRONOMICAL CULTURE IN THE SEVENTEENTH CENTURY

Roberto de Andrade Martins

**Abstract**: There is a remarkable difference between the cultural roles of astronomy in Antiquity and in modern times. Before the scientific revolution, astronomy belonged to an intricate network of knowledges, including religion, mythology, geography, astrology, medicine and other fields. That network was broken after the seventeenth century and astronomy was thereafter regarded as just the study of heavenly phenomena, without much relevance for other fields. The Copernican revolution also entailed a complete change of astronomical terminology, with the loss of terms such as "orb", and the simultaneous substitution by other words, such as "orbit". The terminology changes are analyzed in this paper with the use of Google Books N-gram Viewer. The transformation of astronomical culture that accompanied the Copernican revolution is more substantial than a paradigm shift, because it is not limited to a scientific discipline – it encompasses a whole world view (*Weltanschauung*). This transformation also entailed many losses, as shown in this paper.
**Keywords**: scientific revolution; Copernican revolution; history of astronomy; cultural astronomy; astronomical terminology

## 1. INTRODUCTION

This paper describes the deep changes of the relevance of astronomy to European everyday life accompanying the

MARTINS, Roberto de Andrade. *Studies in History and Philosophy of Science II*. Extrema: Quamcumque Editum, 2021.

Copernican revolution. From Antiquity up to 1600 astronomy was a relevant component of European culture, in several ways. It was taught at the universities both as part of the basic liberal arts and as the technical prerequisite for studying the astrology required in the medical profession. Astronomical knowledge was studied by pilots since it was crucial for oceanic navigation. Astronomy was strongly linked to religious thought and there was a tenuous line of separation between the astronomical heavens and the paradise, in Christian thought. The most common method of measuring time was the sundial, and its construction involved sophisticated knowledge of geocentric astronomy. The apparent motion of the Sun during the year and its connection with the seasons and the variable duration of day and night were well known. Both classical and early modern literature made use of astronomical knowledge. Acquaintance with celestial phenomena and the constellations and their lore was part of culture.

At a time when people spent much time in open spaces and artificial lighting was feeble, knowledge of the apparent motion of the Sun and the study of the shadows it produces, the variable duration of the day and night during the year, the phases of the Moon and other conspicuous phenomena were well known. The motion of the planets, on the other hand, was not described in the oldest extant Greek literature. Notice that the Moon moves around the Earth both in geocentric and heliocentric astronomies; and the apparent motion of the Sun is much easier to understand from the point of view of a geocentric theory, which can deal with all its phenomena.

The situation changed completely during the astronomical revolution. Its emphasis was the study of the motion of the Earth and planetary theory – two subjects with no direct link to everyday life. The heliocentric point of view made it extremely difficult to understand the yearly apparent motion of the Sun and the variation of day and night, or the detailed working of sundials – which became outdated, being replaced by mechanical clocks. Solar tables were still used by the pilots, but their geocentric interpretation was just dropped down – it was not replaced by the heliocentric one. By the end of the seventeenth century, the study of astrology in the universities had declined and the link between the religious heaven and the astronomical sky had been severed.

Any change of paradigm entails both gains and losses, according to Thomas Kuhn. The transformation of astronomical culture that accompanied the Copernican revolution is more substantial than a paradigm shift, because it is not limited to a scientific discipline – it encompasses a whole world view (*Weltanschauung*). This transformation also entailed many losses, as will be shown in this paper.[1]

## 2. ASTRONOMICAL CULTURE IN THE TRANSITION FROM THE 15TH TO THE 16TH CENTURY

In a former paper I studied the situation of astronomical culture in classical Antiquity.[2] The Middle Ages were not dealt with in that article and will not be discussed here, either, because the main target of this paper is the changes that happened during the so-called Copernican astronomical revolution. The importance of astronomy and astrology at the universities, during the late middle ages and Renaissance, can be summarized thus:

> The formal study of both astronomy and astrology in later medieval Europe was firmly based in the universities. Astronomy had by this time become the dominant – and perhaps also the liveliest – portion of the old *quadrivium*, the mathematical portion of the liberal arts, whose subject fields

---

[1] This paper and its first part (the previous essay in this volume) were written for presentation at the conference *Oxford Scientiae 2016: Disciplines of knowing in the early modern world* St Anne's College, University of Oxford, 5-7 July 2016. It was circulated at that time, but it had not been published until now.

[2] MARTINS, Roberto de Andrade. The cultural relevance of astronomy in classical Antiquity, in this volume.

still survived in the diversity of university subjects and texts. In disciplinary terms astrology was widely portrayed, as it had been in antiquity, as an *ars* or *techne*, a practical application of astronomy and astronomical principles. This distinction between the two fields as science and art was maintained more consistently than was the nomenclature; a scholar might use the two terms 'astronomy' and 'astrology' in ways consistent with modern usage, but might also reverse them. Other fields also kept this division between science and art, especially those of the old *quadrivium*. The study of the heavens, then, still formed part of the general studies of philosophy and liberal arts that led to the bachelor's degree. It also held particular and more specialized interest for the medical faculty, because of the fields' perceived abilities to account for celestial influences on health. An expert physician's training and practice included astrology to some degree; the careers of practicing physicians kept astrological practice active in cities and courts. Important courts, in fact, increasingly retained prominent physician/astrologers, a custom established significantly earlier. Astrological practice could also be found at lower social levels, of course, just as other aspects of medical education and practice were not the exclusive province of university-trained experts. (Moyer, 1999, pp. 228-229)

The youth of Nicolaus Copernicus (1473-1543) matched the early phase of the typographical revolution of scientific knowledge. Jérôme de Lalande's restrictive[3] and incomplete *Bibliographie astronomique* lists 160 astronomical publications from that time to the end of the 15th century (Lalande, 1803, pp. 9-29). The correct number is about five times larger.

The presence of astronomical and astrological works among incunabula is significant. In 2015, the *Incunabula Short Title*

---

[3] Lalande did not include in his bibliography popular publications (such as calendars and almanacs) or books on other subjects that contained astronomical information (for instance, medicine books).

*Catalogue* included 27,400 titles[4]. About 16.5% of the incunabula correspond to Klebs' classification of "scientific and medical" (Klebs, 1938), and nearly 20% of the titles analyzed by Klebs correspond to astronomical and astrological works (Klebs, 1932, p. 86; Bühler, 1948).

We can learn a great deal about the astronomical interests in Europe in that period by analyzing the content of those incunabula. The earliest known printed books that could be described as astronomical or astrological were:

1470? – Thomas Aquinas, *De judiciis astrorum*

1472 – Angelus Cato, *De cometa anni 1472*

1472 – Johannes de Sacro Bosco, *Tractatum de spera*

1472 – Johannes de Sacro Bosco, *Spaera mondi*

1472 – Isidore of Seville, *Liber de responsione mundi et astrorum ordinatione*

1472? – Marcus Manilius & Aratus, *Astronomicon; Phaenomena*

1472 – Gerard of Cremona, *Theorica planetarum*

1473 – al-Kabīsī, *Introductorium ad scientiam judicialem astronomiae*

1473 – Aristotle & Averroes, *De caelo et mundo*

1473? – Johannes Regiomontanus, *Kalender*

1474 – Marcus Manilius & Aratus, *Astronomicon; Phaenomena*

1474 – Jean Gerson, *Trilogium astrologiae theologisatae*

1474 – Georg von Peuerbach, *Theoricae novae planetarum*

1474 – Johannes Regiomontanus, *Kalender*

1474 – *Thurecensis physici tractatus de cometis*

1474 – Johannes de Sacro Bosco, *Opusculum spericum*

1474? – Gaietanus de Thienis, *Expositio in libros Aristotelis De coelo et mundo*

1474? – Johannes Regiomontanus, *Disputationes contra Cremonensia in planetarum theoricas deliramenta*

1475 – Johannes Regiomontanus, *Kalender*

---

[4] Most numerical information presented in this section of the paper was obtained by consulting the British Library's *Incunabula Short Title Catalogue*, available online: <http://www.bl.uk/catalogues/istc/index.html>

1475 – Albrecht Kunne, *Calendarium perpetuum*
1475 – Andalo de Nigro, *Opus praeclarissimum astrolabii*
1475 – Jean de Gerson, *Astrologia theologisata*
1476 – Johannes Regiomontanus, *Calendarium*
1476? – Johannes de Sacro Bosco, *Tractatum de spaera*

This list is representative of this period. Notice the presence of authors from classical antiquity (Aritotle, Aratus, Manilius); of some medieval works that were widely used in universities up to the 17th century (Sacrobosco, Gerard of Cremona – later replaced by Peurbach); of ecclesiastical and common calendars; and several astrological works. In the same period we find the publication of non-astronomical works that are relevant for our study of astronomical culture, such as Hesiod's *Works and days*, published in Latin in 1471, and in Greek in 1480. This work was printed 14 times, before 1500. Virgil's *Georgica*, Columella's *De re rustica* and other Roman works on agriculture and farming (by Cato, Varro and others) that make use of astronomical lore, were also printed several times in this period. There were 113 editions of the *Georgics* up to 1500 (the first one in 1469), and 15 editions of Columella's work.

According to George Sarton's census,[5] the twelve most published authors of scientific works in the 15th century were: Albert the Great (151 publications, including any works attributed to him), Aristotle (98 publications, also including spurious works), Hippocrates (52), Arnaldo da Villanova (40), Wenzel Faber (40), Regiomontanus (38), Anianus (37), Jean de Mandeville (36), Canutus (34), Rhazi (34), Bernat de Granollachs (34), Sacrobosco (31) (Sarton, 1938, p. 183). Of these, Albert the Great, Aristotle, Wenzel Faber, Regiomontanus, Anianus, Granollachs and Sacrobosco wrote on astronomy and astrology; some of the other ones included astral information in their books.

---

[5] The numbers cited here are those presented by Sarton. The *Incunabula Short Title Catalogue* identifies a larger number of editions. For information about the *Regimen*, see Ordronaux, 1871.

The most frequently published scientific works in the 15th century (according to Sarton's census) that included astronomical content were: *Regimen Sanitatis* (75 editions), *Liber Aggregationis* ascribed to Albert the Great (61 editions), astrological predictions by Wenzel Faber (38), *Compotus* by Anianus (37), *Lunarium* by Granollachs (34), anonymous prognostications (33), *Sphaera Mundi* by Sacrobosco (31), astrological predictions by Avogario (29), *De Proprietatibus Rerum* by Bartholomew the Englishman (24), astrological predictions by Girolamo Manfredi (20), *Historia Naturalis* by Pliny the Elder (18), *Sfera* by Gregorio – or Leonardo – Dati (17), astrological predictions by Scribanario (17), *Canon Medicinae* by Avicenna (15), *Calendarium* by Regiomontanus (15), *Ephemerides* by Regiomontanus (15), astrological predictions by Novara (14), *De Re Rustica* by Columella (12), *Prognostication* by Lichtenberger (12), astrological predictions by Paulus de Middelburg (11), astrological predictions by Pollich of Mellerstadt (11), *Cosmographia* of Poponius Mela (10), and the anonymous *Compost et Calendrier des Bergers* (10) (Sarton, 1938, pp. 189-191). Notice the large number of highly popular astrological works, in the period. Some other relevant works did not attain the threshold of 10 editions.

Hyginus' *Poetica Astronomica*, the first printed astronomical work presenting many symbolic images of the constellations and of the planets, was published five times before 1500. Of course, Greek and Roman mythology was not acceptable as true, in the Christian context. However, in the middle ages, Dante Alighieri (1265-1321) had provided a new vision of the sacred space of the sky, in his *Divina Commedia*, by populating the heavenly spheres with angels, saints and the Judaico-Christian god. By translating the ancient astral religion into Christian terms, and retaining many of the old astrological ideas, Dante incorporated the continuity of the relationship between the sky and religion (Kay, 1994). Dante's *Divina Commedia* has always been very influential, and it was printed fifteen times before 1500.

Although the massive presence of astrological works is easily perceived, this does not mean that astrology was accepted by everyone – and the same was also true in classical antiquity. Among incunabula we find attacks against astrology by Pierre d'Ailly (*De legibus et sectis contra superstitiosos astronomos*, ca. 1480, ca. 1489), Girolamo Savonarola (*Tractato contra li astrologi*, 1497) and Giovanni Pico della Mirandola (*Compendium sententiarum praeclarissimarum adversus astrologiam*, ca. 1498) – and its defense by Giovanni Abiosi (*Dialogus in astrologiae defensionem*, 1494) and Lucio Bellanti (*Liber de astrologica veritate: et in disputationes Ioannis Pici aduersus astrologos responsiones*, 1498). Most doubts and criticisms of astrology concerned horoscopes (the so-called *judicial astrology*), because the determination of the individual future was sometimes interpreted as contrary to the doctrine of free will (Boas, 1962, p. 168); however, the influence of the stars and planets on the weather, crops, earthquakes, diseases and other physical phenomena was accepted by nearly everyone.

Sarton's analysis of incunabula, based upon Klebs' inventory of scientific and medical incunabula (Klebs, 1938), did not comprise other very popular works, such as the anonymous calendars and almanacs, which always included astronomical content. The earliest known publication of this kind was the *Mainz Kalender*, printed in 1456 or 1457. It was a bleeding and purgation calendar (called *Aderlasskalender* or *Laxierkalendar* in German), which gave details of the lucky and unlucky days on which to bleed or take medicine in a given year (Green, 2012, p. 29; Luke, 1989, p. 56). Printed in a single paper sheet, those medical calendars were very popular. At least 46 were printed before 1480, and more than one hundred up to 1500 (Berry & Poole, 1966, p. 13; Sudhoff, 1908, pp. 261-432). Those medical calendars introduced for the first time, in printing, the famous images of the cosmic man (or woman) with the relation between the Zodiac and the parts of the body. If we include all types of calendars and almanacs published up to the end of the 15th century, their number exceeds 500. Almanacs were published both in Europe and in the New World (Guerra, 1961).

Islamic medicine, that was highly influential in Europe and well represented in books published during this period, accepted the humoral theory of Hippocrates, developed by Galen, its association with the Aristotelian theory of the four elements and four qualities, together with the theory of individual temperaments, influenced by the time of birth, and the consideration of man as a microcosm. The seven cervical and the twelve dorsal vertebrae corresponded to the seven planets and twelve signs of the Zodiac, as well as the days of the week and the months of the year; and the total number of discs of the vertebrae, which they considered to be 28, to the stations of the Moon. The correspondence and "sympathy" between various orders of cosmic reality form the philosophical background of Islamic medicine (Nasr, 1968, pp. 219-224).

Astrology had become highly influential in medieval Europe after the introduction of Islamic works, such as those translated under Alfonso X (Harvey, 1977; Rodríguez, 2007). The level of popularity of astrology during that period can be recognized, for instance, in the heavy use of astrological references in Chaucher's *Canterbury Tales* (Wood, 1970; Cartwright, 2005). The most influential Islamic astronomical/astrological authors whose books were printed in the 15th century were Alchabitius (1473, 1485, 1491), Albohazen (1485), Abdilaziz (1482, 1485), Albumasar (1488, 1489), Messahalla (1493), Almanzor (1493).

Ptolemy's *Almagest* was not printed during the 15th century, but his geographical work (*Cosmographia*) was printed seven times, and his astrological treatise (*Quadripartitum*, or *Tetrabiblos*) had two editions. There were also six editions of Strabo's *Geographia*. Regiomontanus's abridgment of the *Almagest* (*Epitoma in Almagestum Ptolemaei*) was first printed in 1496. At the universities, up to the 17th century, the study of astronomy followed Sacrobosco's *Tractatus de Sphaera* (and commentaries) for the general description of the world (Gingerich, 1988), complemented by the *Theorica Planetarum* of Gerard of Cremona or the *Theorica Nova* of Peurbach, that supplied the technical discussion of the motion of the planets (Pedersen, 1981). Astrologers usually relied on tables such as those of Regiomontanus' *Ephemerides* for numerical data, instead of calculating the locations of the planets (Boas, 1962, p. 169).

According to Seyyed Hossein Nasr, in medieval Islam, alchemy and astrology were described as complementary. Although astrology describes the direct celestial influences, alchemy deals with substances of the sublunary world that were generated under astral influences and that are directly related to the planets. The seven planets were represented by seven metals, being represented by the same symbols as a part of the wide astral symbolic network: lead (Saturn), tin (Jupiter), iron (Mars), gold (Sun), copper (Venus), quicksilver (Mercury), silver (Moon) (Nasr, 1968, pp. 250-251). The astral influence upon terrestrial materials was also the basis of the wide development of talismans, or *Scientia imaginum* developed by Islamic authors and that was treated, for instance, in the books *De mineralibus* and *Speculum astronomiae* ascribed to Albertus Magnus (Weill-Parot, 1999). Hieronymus Torrella's *Opus praeclarum de imaginibus astrologicis* (1496) and other later works were dedicated to this subject.

The study of geography and the sea navigations were still strongly dependent on astronomy. The observation of the stars or the Sun, using an astrolabe, was the only available method for determining the geographical latitude. However, although the Portuguese and Spanish long distance voyages were already starting, we find no publication devoted to astronomical navigation in the late 15th century. It is true that Abraham ben Samuel Zacuto and José Vizinho authored the *Almanach perpetuum* first published in Portugal in 1496 and that this work was used by Portuguese pilots; explicit manuals for astronomical navigation were only published in the 16th century, by Fernandez de Enciso, Francisco Faleiro, Pedro de Medina, Martin Cortes and others (Proverbio, 1994). We should point out that, in the 16th century, the epic poem *Lusíadas*, by Luís de Camões (1524-1580), describing the Portuguese voyages and conquests, was full of astronomical references (Silva, 1915). This shows both the relevance of astronomy for the long distance travels and the acquaintance of educated persons of the period with astronomical knowledge.

Since the 12th century, contrary to the situation in antiquity, European pilots were acquainted and used the magnetic compass for the determination of geographical directions (Smith, 1992). However, this instrument did not replace the astronomical means of orientation. First, it is relevant to point out that Islamic authors had associated the orientation of magnetic bodies to an astral effect (Weill-Parot, 2015). Up to Gilbert's 1600 work on the magnet, the compass was not a replacement for astronomical orientation: it was a special instrument acted by the stars. Most authors accepted that the magnetic needle pointed to the celestial pole. Besides that, in the 16th century pilots perceived that the magnetic compass was not reliable (at some places it had a large deviation from the astronomical meridian) and only the Sun and the stars provided the true north-south direction (Mitchell, 1937).

Hence, in the early 16th century, astronomy and astrology were relevant components of the European culture. They were taught at the universities; many publications on those subjects – both technical and popular ones – had a wide circulation; astrology and astronomy were the basis of medical and agricultural practices; the stars, the Sun and the Moon remained the basic references for the measurement of time, the establishment of geographical data and the prediction of weather; calendars and almanacs established the adequate days for several activities – including religious ones; navigation was guided by sky observations; astronomy and astrology references frequently appeared in literature; astral mythology and celestial symbolism was still influential; and belief of the influence of the heavens upon matter was the basis of talismanic magic and part of the alchemical theory of that time.

## 3. THE COPERNICAN REVOLUTION AND THE CHANGES OF ASTRONOMICAL CULTURE

The so-called astronomical revolution started with the publication of Nicolaus Copernicus' *De revolutionibus orbium coeslestium* in 1543, received important contributions from Giordano Bruno, Johannes Kepler, Tycho Brahe, William Gilbert, Galileo Galilei, René Descartes and several other authors, and concluded in the middle of the 18th century, with the general acceptance of Isaac Newton's mechanics and theory of gravitation. The change was already complete when the *Encyclopédie ou dictionnaire raisonné des sciences, des arts et des métiers* was published, in 1745-1756 (Kuhn, 1957; Koyré, 1961; Boas Hall, 1962; Rupert Hall, 1983; Gingerich, 2004).

It is impossible to describe this revolution in detail, here. Copernicus' main contribution was a change from geocentrism to a heliocentric theory of the motion of the planets; however, from the middle of the 16th century to the beginning of the 18th century there were several other fundamental contributions that provided a complete cosmological upheaval. The rational acceptance of the motion of the Earth around the Sun required a new mechanics, with the abandonment of the distinction of natural and violent motions and a novel interpretation of gravity. The whole theory of the five elements had to be abandoned, since there was no essential difference between the matter around us and the celestial bodies. Although Copernicus still accepted the celestial spheres and tried to explain the motion of the planets using a combination of circular motions, those principles were gradually dropped out, being ultimately replaced by the idea of celestial bodies moving in empty space (or in a very rarefied ether) according to the laws of Newtonian mechanics and acted upon by gravitational forces. The sphere of the stars was abandoned, and those celestial bodies were regarded as hanging in infinite (or indefinite) space at widely different distances of the Earth. Although the Sun became the center of the planetary system, the universe had no central point.

It became conceivable that there were other worlds similar to the Earth, with living beings, rotating around some of the stars.

The history of the Copernican revolution has usually being studied by analyzing the contributions and opinions of a small number of relevant individuals. In recent years, it became possible to apply new computational tools to the study of history, allowing us to have a different view of collective cultural changes, through the analysis of thousands (or millions) of publications. One of those recent tools is the *Google Books Ngram Viewer* (Michel *et al.*, 2011)[6]. It has been described, discussed and applied to historical research by several authors (wee, for instance: Yeung & Jatowt, 2011; Gibbs & Cohen, 2011; Virues-Ortega & Pear, 2012; Greenfield, 2013; Solée et al., 2013; Genovese, 2015; Milligan, 2012).

**Fig. 1.** *Google Books Ngram Viewer* frequency of appearance (ten years smoothing) of the words "orb" and "orbs" in English books, relative to the appearance of the words "star", "planet" (and their plurals), "astronomy", "astronomical" and "astronomic".[7]

---

[6] Available at <http://books.google.com/ngrams>.

[7] All graphs presented in this paper were obtained in March 2016, when this essay was prepared.

Using the *Ngram Viewer* it is possible to perceive, for instance, the fast drop of the use of "orb" (or "orbs") in the astronomical context, in the 17th century (Fig. 1). Up to that time, this word had been used to describe the invisible celestial spheres that held and were responsible for the motions of the stars and planets. Shortly before 1700, it had almost disappeared, due to the ongoing astronomical revolution.

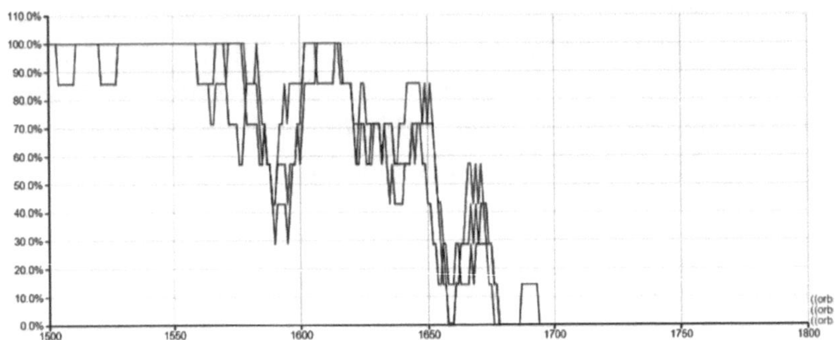

**Fig. 2**. *Google Books Ngram Viewer* frequency of appearance (three years smoothing) of the phrases "orb of Mars" and "sphere of Mars" in English books, relative to the appearance of the word "Mars", and corresponding frequencies for Venus and Jupiter. The main trend does not depend on the planet.

Up to 1600, the motion of the planets was described as produced by their orbs. In the 17th century, phrases such as "orb of Venus" or "sphere of Venus" vanished (Fig. 2)[8]. A similar trend can be noticed regarding the phrases "celestial sphere" and "heavenly sphere" (and their plurals), that suffered a sudden drop in the 17th century and had almost disappeared in the early 18th century – around 1720 or 1730 (Fig. 3).

Until the beginning of the 17th century, the Sun was supposed to be carried around the Earth by its deferent or

---

[8] Unfortunately, the statistics provided by the *Google Books Ngram Viewer* for English books prior to 1600 is poor.

eccentric circle. Around 1650, the phrases "deferent of the Sun" and "eccentric of the Sun" disappeared (Fig. 4). Likewise, the words epicycle, eccentric and deferent that were used to describe the motion of the planets practically vanished around 1680, reappearing, however, in historical accounts after that time (Fig. 5).

**Fig. 3**. *Google Books Ngram Viewer* frequency of appearance (ten years smoothing) of the phrases "celestial sphere" and "heavenly sphere" (and their plurals) in English books, relative to the appearance of the word "planets" (lower line), and relative to the appearance of the word "stars" (upper line).

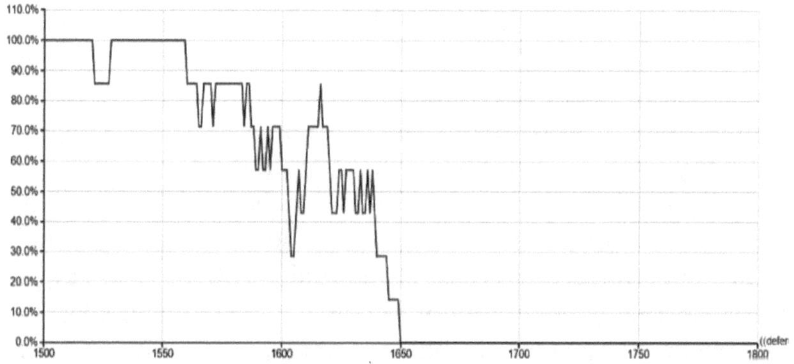

**Fig. 4**. *Google Books Ngram Viewer* frequency of appearance (three years smoothing) of the phrases "deferent of the Sun" or "eccentric of the Sun" in English books, relative to the appearance of the word "Sun".

189

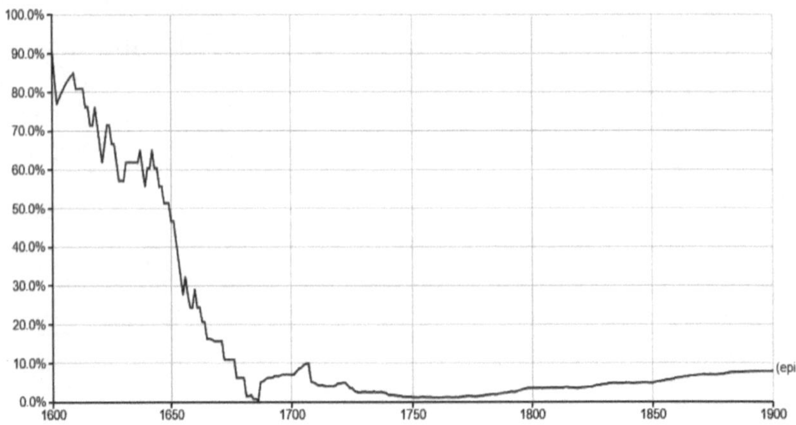


**Fig. 5**. *Google Books Ngram Viewer* frequency of appearance (ten years smoothing) of the words epicycle, eccentric or deferent  in English books, relative to the appearance of the words "planet", "planets", "astronomy", "astronomical" and "astronomic".

Those statistical results confirm what one would expect from the current view of the Copernican revolution. Observe, however, that the analysis of different words of phrases provide dissimilar extinction curves and distinct disappearance dates. Notice, also, that some specific terms (such as "orb") die very hard (Fig. 1).

In the same period, it is possible to discern some changes that were not directly produced by the Copernican revolution, although they accompanied it. Up to the end of the 16th century, sundials were the most common instrument for measuring time, and they were described in astronomical books. The improvement and dissemination of mechanical clocks – and, especially, pendulum clocks – during the 17th century entailed the disappearance of interest in sundials around 1730 (Fig. 6). Of course, the development of the new mechanical clocks could have occurred one century before the Copernican revolution; it is just a contingency that the two historical trends occurred at about the same period.

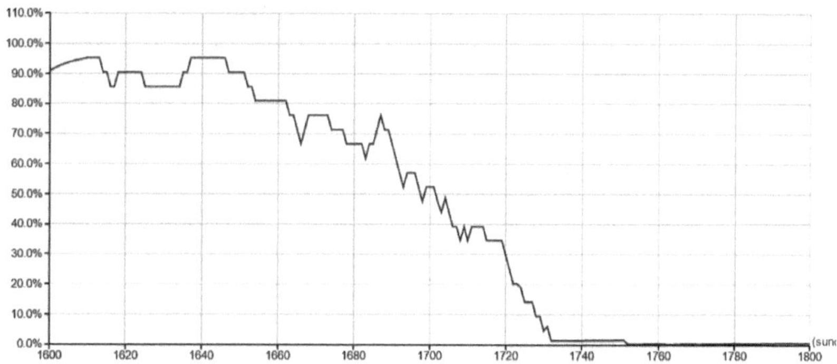

**Fig. 6**. *Google Books Ngram Viewer* frequency of appearance (ten years smoothing) of the word "sundial" (or "sun dial", or "sundials") in English books, relative to the appearance of the words "astronomy" or "astronomical" or "astronomic".

With the abandonment of the sundial, the division of the bright part of the day in 12 parts was abandoned, and the very concepts of "artificial hour" and "natural hour" disappeared from the astronomical literature around 1730 (Fig. 7).

**Fig. 7**. *Google Books Ngram Viewer* frequency of appearance (ten years smoothing) of the phrases "natural hours" or "artificial hours" in English books, relative to the appearance of the words "astronomy" or "astronomical" or "astronomic".

From Antiquity to the Renaissance, the determination of the periods of the year and its cyclical phenomena were associated to the observation of the rising and setting of the stars. However, the very *description* of helical, cosmical and acronychal rising and setting of the constellations disappeared from the astronomical literature around 1730, accompanying the Copernican revolution (Fig. 8).

**Fig. 8**. *Google Books Ngram Viewer* frequency of appearance (ten years smoothing) of phrases containing helical or cosmical or acronychal and rising or setting in English books, relative to the appearance of the words "astronomy" or "astronomical" or "astronomic".

Besides the constellations of the Zodiac, there were other constellations or stars that had been relevant signposts of the divisions of the year, since Antiquity: the Pleiades, Sirius (or Dog Star), Orion, Arcturus, Bootes, the Hyades. References to those stars did not disappear, but reached a small level around 1680, and after a rise around 1710, they collapsed to a very low level after 1750 (Fig. 9). A very similar trend can be observed regarding the presence of the classical names of the winds (Boreas, Notos, Zephyros, Euros), which had been associated to the seasons and to astronomical phenomena since Antiquity. The frequency of occurrence of those names becomes small around 1650-1690, has an increase around 1700, and drops to negligible frequencies after 1750 (Fig. 10).

**Fig. 9**. *Google Books Ngram Viewer* frequency of appearance (ten years smoothing) of the names of some common stars and constellations (Pleiades, Sirius or Dog Star, Orion, Arcturus, Bootes, Hyades) in English books, relative to the appearance of the words "astronomy" or "astronomical" or "astronomic".

**Fig. 10**. *Google Books Ngram Viewer* frequency of appearance (ten years smoothing) of the traditional names of the four main winds (Boreas, Notos or Notus, Zephyros or Zephyrys, Euros or Eurus) in English books, relative to the appearance of the words "astronomy" or "astronomical" or "astronomic".

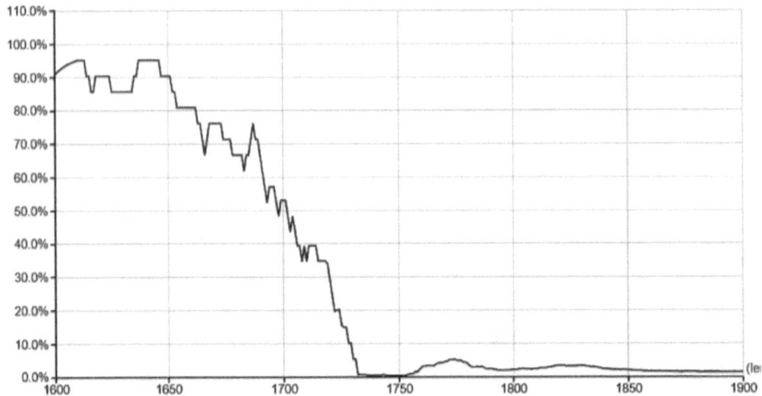

**Fig. 11**. *Google Books Ngram Viewer* frequency of appearance (ten years smoothing) of the phrases "length of the day" or "duration of the day" (or night) in English books, relative to the appearance of the words "astronomy" or "astronomical" or "astronomic".

The variation of the lengths of day and night during the year and in different places of the Earth was an important phenomenon presented in most astronomical works before the Copernican revolution. Unexpectedly, this subject virtually disappeared from the books around 1730 (Fig. 11). The *explanation* of the effect is, of course, different in the geocentric and in the heliocentric theories; but the phenomenon itself is relevant and observable, and there was no evident reason for dismissing its study in the 18th century.

From the Middle Ages to the Renaissance, the most popular astronomical textbook was Sacrobosco's *Tractatus de Sphaera*, together with its commentaries and complements. In the beginning of the 17th century Sacrobosco was still highly cited, but around 1680 his name virtually vanished from the literature (Fig. 12). It was never replaced by any other equally popular astronomical textbook.

Interest in astrology, as shown by the frequency of appearance of this word, decreased in the 17th century, suffered a strong revival in the early 18th century and was still sizeable in the early 19th century (Fig. 13). It is clear that the Copernican

194

revolution did not "kill" astrology. The causes behind the decreasing interest in astrology during this period are still open to debate, although many different explanations have been suggested (Ferreira, 2005).

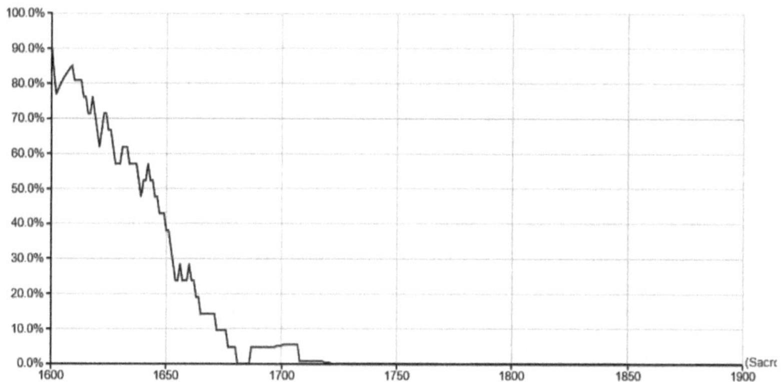

**Fig. 12**. *Google Books Ngram Viewer* frequency of appearance (ten years smoothing) of the name Sacrobosto (with the variants Sacrobusto and Sacro Bosco) in English books, relative to the appearance of the words "astronomy" or "astronomical" or "astronomic".

**Fig. 13**. *Google Books Ngram Viewer* frequency of appearance (ten years smoothing) of the words "astrology" or "astrological" or "astrologic" in English books, relative to the appearance of the words "astronomy" or "astronomical" or "astronomic".

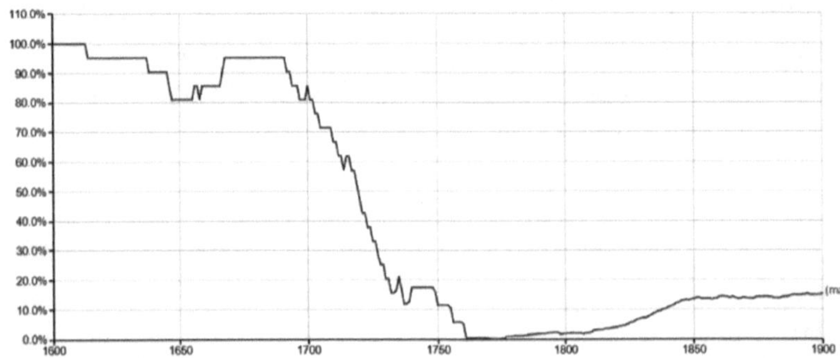

**Fig. 14**. *Google Books Ngram Viewer* frequency of appearance (ten years smoothing) of the phrase "magnetic influence" in English books, relative to the appearance of the words "astrology" or "astrological" or "astrologic".

Before the scientific revolution, the concept of astrological effects was usually linked to the notion of magnetic influences – an idea that had been reinforced by the discovery of the directional property of magnets in the Middle Ages. Even after Gilbert's work on magnetism, this idea was still highly influential, but nearly disappeared around 1760 – reappearing later, with other magnetic beliefs, such as Mesmerism (Fig. 14).

**Fig. 15**. *Google Books Ngram Viewer* frequency of appearance (ten years smoothing) of the phrase "critical day" and its plural, in English books, relative to the appearance of the word "medicine".

The traditional astrological medicine also declined, during the 17th and 18th centuries. This can be noticed, for instance, by the decrease of the presence of the phrase "critical day" (and its plural). After frequency oscillations during the 17th century, this expression decreased markedly during the early 18th century and vanished around 1750 (Fig. 15).

The statistical data used in the analysis presented above has some shortcomings. The *Google Books Ngram Viewer* does not include books in Latin, which was a relevant scientific language during that period. Only English books were used for producing the graphs shown in this paper; different results arise for other languages. Besides that, the database does not accept the full use of Boolean operators. Even with those limitations, this tool provides relevant results. Of course, the data provided by this instrument does not offer *explanations*, but it supplies relevant information that requires historical interpretation.

## 4. FINAL COMMENTS

In classical antiquity, astronomy had a strong connection with everyday life and was deeply embedded in the philosophical and religious culture. Although this relationship changed during the Middle Ages, due to both Islamic and Christian influences, astronomy remained of paramount importance in the culture and practical life until Renaissance.

During the period of the Copernican revolution, it is possible to notice the decline of astrology and of all beliefs concerning the celestial influence on health, weather and other phenomena – although almanacs and other popular publications were still published and read. Some explanations can be offered for this change. According to the new cosmology, the planets could not be interpreted as special and almost supernatural bodies: they were just big rocks; the stars (including the Sun) were just big fiery material bodies. They did not emit any special rays – only light and heat. They did not act upon other bodies by a mysterious magnetic influence – they just had a gravitational power that was unable to produce any strong physical effect on terrestrial bodies, except in the case of some special phenomena, such as the tides. The constellations were not real groups of stars with special powers: they only appeared to be near to one another when observed from a particular point of space – our position. The motion of the planets could be explained using physics and mathematics, without any additional assumption about intelligent beings (such as gods or angels) associated to them. The concept of a finite universe with an outer spherical shell was discarded, leaving no place in the astronomical scheme for God, saints or angels. The sky lost its mysterious and sacred status and became just a set of material bodies moving under physical laws.

So, this new *Weltanschauung* may account for several differences between the old and new astronomical and astrological cultures. Other changes, although they occurred at the same time, require a different explanation.

Time became a completely artificial device, on all scales. The month was not related anymore to the phases of the Moon, the beginning of the year and its important periods were not related to the rise and setting of the stars. The religious festivals were not explicitly related to astronomical events, anymore. Hours were now measured by a mechanical device, that seemed to have no relation whatever to the sky and astronomical phenomena. The astronomical division of the time between successive midnights in 24 parts replaced the observation of the rising and setting of the Sun and the use of different hours in each period of the year. The fundamental phenomenon of the variation of the length of day and night was not studied any more. The neglect of this last astronomical topic may have a specific explanation. In the geocentric theory, it was very easy to understand how the motion of the Sun around the Earth produced this variation of the duration of day and night, and its relation to the latitude of the observer. The analysis was not quite simple in the heliocentric theory, and perhaps for that reason the astronomical books left this phenomenon out of their scope.

The geographical directions were related to the structure of the celestial sphere and its motion. The heavenly poles were directly observable. Once the sky became a set of detached stars and there was no sphere rotating around the Earth, there were no celestial poles, according to the original meaning of this word. The Earth was supposed to rotate and to have poles, but they were not observable. Accordingly, most people began to attach an absolute meaning to the magnetic meridian, as if it were the original and meaningful definition of the north-south direction. Only pilots and specialists knew that the magnetic compass could suffer deviations, and they kept the astronomical meaning of the main geographical directions. The popular awareness of the relation between the four main directions and the structure of the sky vanished.

From Antiquity to the Renaissance, the popular astronomical knowledge was strongly associated to observations. Everyone could perceive the motion of the Sun, of the stars, of the Moon. After the Copernican revolution, direct observations were not meaningful anymore, because the Sun and the stars did not move. A scientific knowledge that had direct connection to the observable world around us was replaced by an abstract knowledge that had no relation to the visible world, for the general population. We do not observe the motion of the Earth; we do observe the motion of the Sun; but we are mistaken, only the astronomers know the truth, by some mysterious methods. Of course, everybody knew that naked eye observations had been improved and replaced by the use of the telescope. As a matter of fact, the telescope does not show that the Earth moves and the Sun is still – but the general public thought (and still thinks) that astronomers can *see* that this is true.

In the 18th century, the study of the sky became a highly specialized knowledge, and the astronomical culture of the general population decreased. The losses produced by the Copernican revolution were significant, and they deserve more attention by historians.

## ACKNOWLEDGMENTS

The author is grateful for the support he received from the São Paulo State Foundation for the Support of Research (FAPESP) and from the Brazilian Council for Scientific and Technological Development (CNPq) for the elaboration of this research.

## BIBLIOGRAPHICAL REFERENCES

BERRY, William Turner; POOLE, Herbert Edmund. *Annals of printing: a chronological encyclopaedia from the earliest times to 1950*. London: Blandford Press, 1966.

BOAS, Marie. *The scientific renaissance, 1450-1630*. New York: Harper & Brothers, 1962.

BÜHLER, Curt F. The statistics of scientific incunabula. *Isis*, 39 (3): 163-168, 1948.

CARTWRIGHT, John. Medieval cosmology and European literature: Dante and Chaucer. Pp. 1-30, in: CARTWRIGHT, John H.; BAKER, Brian (eds.). *Literature and science: social impact and interaction*. Santa Barbara: ABD-Clio, 2005.

FERREIRA, Juliana Mesquita Hidalgo. *As influências celestes e a Revolução Científica: a astrologia em debate na Inglaterra do século XVII*. PhD Dissertation in Philosophy. São Paulo: Pontifícia Universidade Católica de São Paulo, 2005.

GENOVESE, Jeremy E. C. Interest in astrology and phrenology over two centuries: a Google Ngram study. *Psychological Reports*, 117: 940-943, 2015.

GIBBS, Frederick W.; COHEN, Daniel J. A conversation with data: prospecting Victorian words and ideas. *Victorian Studies*, 54 (1): 69-77, 2011.

GINGERICH, Owen. Sacrobosco as a textbook. *Journal for the History of Astronomy*, 19: 269-273, 1988.

GINGERICH, Owen. *The book nobody read: chasing the revolutions of Nicolaus Copernicus*. New York: Walker, 2004.

GREEN, Jonathan. *Printing and prophecy: prognostication and media change 1450-1550*. Ann Arbor: University of Michigan Press, 2012.

GREENFIELD, Patricia M. The changing psychology of culture from 1800 through 2000. *Psychological Science*, 24 (9): 1722-1731, 2013.

GROSCHWITZ, Helmut. *Mondzeiten. Zu Genese und Praxis moderner Mondkalender*. Münster: Waxmann Verlag, 2008.

GUERRA, Francisco. Medical almanacs of the American colonial period. *Journal of the History of Medicine and Allied Sciences*, 16 (3): 234-255, 1961.

HALL, Alfred Rupert. *The revolution in science, 1500-1750*. London: Longman, 1983.

HALL, Marie Boas. *The scientific Renaissance, 1450-1630*. New York: Harper, 1962.

HARVEY, Leonard Patrick. The Alfonsine school of translators: Translations from Arabic into Castilian produced under the patronage of Alfonso the Wise of Castile (1221–1252–1284). *Journal of the Royal Asiatic Society of Great Britain & Ireland* (new series), 1: 109-117, 1977.

KAY, Richard. *Dante's Christian astrology*. Philadelphia: University of Pennsylvania Press, 1994.

KLEBS, Arnold C. Gleanings from incunabula of science and medicine. *Papers of the Bibliographical Society of America*, 26 (1/2): 52-88, 1932.

KLEBS, Arnold C. Incunabula scientifica et medica. *Osiris*, 4: 1-359, 1938.

KOYRÉ, Alexandre. *La révolution astronomique : Copernic, Kepler, Borelli*. Paris: Hermann, 1961.

KUHN, Thomas S. The Copernican revolution: planetary astronomy in the development of western thought. Cambridge, MA: Harvard University Press, 1957.

LALANDE, Jérôme de. *Bibliographie astronomique; avec l'histoire de l'astronomie depuis 1781 jusqu'à 1802*. Paris: De l'Imprimérie de la République, 1803.

LUKE, Carmen. *Pedagogy, printing and Protestantism*. Albany: State University of New York Press, 1989.

MICHEL, Jean-Baptiste: SHEN, Yuan Kui; AIDEN, Aviva Presser; VERES, Adrian; GRAY, Matthew K.; The Google Books Team; PICKETT, Joseph P.; HOIBERG, Dale ;CLANCY, Dan; NORVIG, Peter; ORWANT, Jon; PINKER, Steven; NOWAK, Martin A.; AIDEN, Erez Lieberman. Quantitative analysis of culture using millions of digitized books. *Science*, 331 (6014): 176-182, 2011.

MILLIGAN, Ian. Mining the 'Internet graveyard': rethinking the historians' toolkit. *Journal of the Canadian Historical Association*, 23 (2): 21-64, 2012.

MITCHELL, Alexander Crichton. Chapters in the history of terrestrial magnetism. II. The discovery of declination. *Terrestrial Magnetism and Atmospheric Electricity*, 42 (3): 241-280, 1937.

MOYER, Ann. The astronomers' game: astrology and university culture in the fifteenth and sixteenth centuries. *Early Science and Medicine*, 4 (3): 228-250, 1999.

NASR, Seyyed Hossein. *Science and civilization in Islam*. Cambridge, MA: Harvard University Press, 1968.

ORDRONAUX, John. *Regimen sanitatis Salernitanum. Code of health of the school of Salernum*. Philadelphia: J. B. Lippincott, 1871.

PEDERSEN, Olaf. The origins of the Theorica Planetarum. *Journal for the History of Astronomy*, 12: 113-123, 1981.

PROVERBIO, Edoardo. Astronomical and sailing tables from the second half of the XVth century to the middle of the XVIth century. *Memorie della Società Astronomica Italiana*, 65: 469-496, 1994.

RODRÍGUEZ, Ana Domínguez. Astrología y mitología en los manuscritos ilustrados de Alfonso X El Sabio. *En la España Medieval*, 30: 27-64, 2007.

SARTON, George. The scientific literature transmitted through the incunabula. *Osiris*, 5: 41-245, 1938.

SILVA, Luciano Pereira da. *A astronomia dos Lusíadas*. Coimbra: Imprensa da Universidade de Coimbra, 1915.

SMITH. Julian A. Precursors to Peregrinus: The early history of magnetism and the mariner's compass in Europe. *Journal of Medieval History*, 18 (1): 21-74, 1992.

SOLÉE, Ricard V.; VALVERDE, Sergi; CASALS, Marti Rosas; KAUFFMAN, Stuart; FARMER, Doyne; ELDREDGE, Niles. The evolutionary ecology of technological innovations. *Complexity*, 18, (4): 15-27, 2013.

SUDHOFF, Karl. *Deutsche medizinische Inkunabeln; bibliographisch-literarische Untersuchungen*. Leipzig: Johann Ambrosius Barth, 1908.

VIRUES-ORTEGA, Javier; PEAR, Joseph J. A history of "behavior" and "mind": use of behavioral and cognitive terms in the 20th century. *The Psychological Record*, 65 (1): 23-30, 2015.

WEILL-PAROT, Nicolas. Causalité astrale et "science des images" au Moyen Age: éléments de réflexion. *Revue d'Histoire des Sciences*, 52 (2): 207-240, 1999.

WEILL-PAROT, Nicolas. L'attraction magnétique entre influence astrale et astrologie en Moyen Âge (XIIIe-XVe siècle). *Philosophical Readings*, 7 (1): 55-70, 2015.

WOOD, Chauncy. *Chaucer and the country of the stars: poetic uses of astrological imagery*. Princeton: Princeton University Press, 1970.

YEUNG, Ching-Man Au; JATOWT, Adam. Studying how the past is remembered: towards computational history through large scale text mining. Pp. 1231-1240, in: *Proceedings of the 20th ACM International Conference on Information and*

*Knowledge Management*. New York: Association for Computing Machinery / ACM, 2011.